COGNITIVE STRUCTURE AND CONCEPTUAL CHANGE

EDUCATIONAL PSYCHOLOGY

Allen J. Edwards, Series Editor
Department of Psychology
Southwest Missouri State University
Springfield, Missouri

In Preparation

Gilbert R. Austin and Herbert Garber (eds.). Research on Exemplary Schools.

Published

Leo H. T. West and A. Leon Pines (eds.). Cognitive Structure and Conceptual Change

Catherine T. Best (ed.). Hemispheric Function and Collaboration in the Child

Charles D. Holley and Donald F. Dansereau (eds.). Spatial Learning Strategies: Techniques, Applications, and Related Issues

John R. Kirby (ed.). Cognitive Strategies and Educational Performance

Penelope L. Peterson, Louise C. Wilkinson, and Maureen Hallinan (eds.). The Social Context of Instruction: Group Organization and Group Processes

Michael J. A. Howe (ed.). Learning from Television: Psychological and Educational Research

Ursula Kirk (ed.). Neuropsychology of Language, Reading, and Spelling

Judith Worell (ed.). Psychological Development in the Elementary Years

Wayne Otto and Sandra White (eds.). Reading Expository Material

John B. Biggs and Kevin F. Collis. Evaluating the Quality of Learning: The Solo Taxonomy (Structure of the Observed Learning Outcome)

Gilbert R. Austin and Herbert Garber (eds.). The Rise and Fall of National Test Scores

Lynne Feagans and Dale C. Farran (eds.). The Language of Children Reared in Poverty: Implications for Evaluation and Intervention

Patricia A. Schmuck, W. W. Charters, Jr., and Richard O. Carlson (eds.). Educational Policy and Management: Sex Differentials

Phillip S. Strain and Mary Margaret Kerr. Mainstreaming of Children in Schools: Research and Programmatic Issues

The list of titles in this series continues at the end of this volume.

COGNITIVE STRUCTURE AND CONCEPTUAL CHANGE

Edited by

LEO H. T. WEST
Higher Education Advisory and Research Unit
Monash University
Clayton
Victoria, Australia

A. LEON PINES
Visiting Scholar
School of Education
University of California, Berkeley

1985

ACADEMIC PRESS, INC.

(Harcourt Brace Jovanovich, Publishers)

Orlando   San Diego   New York   London
Toronto   Montreal   Sydney   Tokyo

ACADEMIC PRESS, INC.
Orlando, Florida 32887

United Kingdom Edition published by
ACADEMIC PRESS INC. (LONDON) LTD.
24–28 Oval Road, London NW1 7DX

Library of Congress Cataloging in Publication Data

Main entry under title:

Cognitive structure and conceptual change.

   (Educational psychology series)
   Includes index.
   1. Cognition. 2. Change (Psychology) 3. Learning,
Psychology of. 4. Science--Study and teaching, I. West,
Leo H. T. II. Pines, A. Leon. III. Series.
BF311.C5534   1985   153   84-20415
ISBN 0-12-744590-0 (alk. paper)

PRINTED IN THE UNITED STATES OF AMERICA

85 86 87 88   9 8 7 6 5 4 3 2 1

For Misha
You were there for me.

A. L. P.

CONTENTS

II
STABILITY AND CHANGE IN CONCEPTUAL UNDERSTANDING

Contributors

Numbers in parentheses indicate the pages on which the authors' contributions begin.

AUDREY B. CHAMPAGNE (61, 163), Learning Research and Development Center, University of Pittsburgh, Pittsburgh, Pennsylvania 15260

PETER J. FENSHAM (29), Faculty of Education, Monash University, Clayton, Victoria 3168, Australia

JANICE E. GARRARD (29), Faculty of Education, Monash University, Clayton, Victoria 3168, Australia

JOHN K. GILBERT (11), Department of Educational Studies, Surrey University, Guildford GU2 5XH, Surrey, England

RICHARD F. GUNSTONE (61, 163), Faculty of Education, Monash University, Clayton, Victoria 3168, Australia

JOHN O. HEAD (91), Centre for Science Education, Chelsea College, Bridges Place, London SW6 4HR, England

MARIANA G. A'B. HEWSON[1] (153), National Institute for Personnel Research, Braamfontein 2017, South Africa

BENGT JOHANSSON (233), Department of Education, University of Göteborg, Mölndal, Sweden

LEOPOLD E. KLOPFER (61, 163), Learning Research and Development Center, University of Pittsburgh, Pittsburgh, Pennsylvania 15260

FERENCE MARTON (233), Department of Educational Research, University of Göteborg, Mölndal, Sweden

JOSEPH D. NOVAK (189), Department of Education, Cornell University, Ithaca, New York 14853

ROGER J. OSBORNE (11), Science Education Research Unit, University of Waikato, Hamilton, New Zealand

A. LEON PINES (1, 101), School of Education, University of California, Berkeley, Berkeley, California 94720

GEORGE J. POSNER (211), Department of Education, Cornell University, Ithaca, New York 14853

[1]Present address: Center for Educational Research, University of Wisconsin at Madison, Madison, Wisconsin 53706.

F. REIF (133), Physics Department and School of Education, University of California, Berkeley, Berkeley, California 94720

THOMAS J. SHUELL (117), Department of Counseling and Educational Psychology, State University of New York at Buffalo, Buffalo, New York 14260

KENNETH A. STRIKE (211), Department of Education, Cornell University, Ithaca, New York 14853

CLIVE R. SUTTON (91), University of Leicester, Leicester LE1 7RH, England

LENNART SVENSSON (233), Department of Education, University of Göteborg, Mölndal, Sweden

D. MICHAEL WATTS (11), Secondary Science Curriculum Review, London W1N 6BA, England

LEO H. T. WEST (1, 29), Higher Education Advisory and Research Unit, Monash University, Clayton, Victoria 3168, Australia

RICHARD T. WHITE (51), Faculty of Education, Monash University, Clayton, Victoria 3168, Australia

M. C. WITTROCK (259), Department of Education, University of California, Los Angeles, Los Angeles, California 90024

Cognitive Structure and Conceptual Change is part of a new era in educational research and practice. Research in naturalistic settings using qualitative methods is now recognized as the most appropriate approach in many situations. The application of these methodologies to student learning has revolutionized our understanding of the learning process. This volume is in the vanguard of that revolution.

The book is divided into two major parts: The first deals almost exclusively with attempts to elucidate cognitive structure; the second moves toward describing ways of changing cognitive structure. None of the material presented is a rehash of existing work in educational theory, research, and practice; rather, each chapter breaks new ground.

This book will be useful to all educators—theoreticians, researchers, and practitioners. Educators in universities—both those in education and those in specific subject-matter areas—will find the book and the ideas presented in it useful in their teaching. For example, specific subject-matter areas in physics, chemistry, and related disciplines are used in both the first and the second parts of the book and can serve as a guide for those who wish to teach and research teaching in these areas. For schools of education, in which all too often those "awful methods courses" (to use James Bryant Conant's phrase) predominate, but which wish to improve, this book is a sine qua non. Instructional theories must be ensconsed within a coherent theoretical framework and erected upon a firm empirical base. This book can serve as a text and reference book for graduate students in education and for all other related disciplines. Teachers will benefit from this book because its methods and theories refer to real learning in real classroom settings.

We wish to thank formally those individuals who have contributed to the book. The authors whose excellent chapters form the basis of the volume were set difficult deadlines; their cooperation and patience are greatly appreciated.

A number of the chapters were originally commissioned for two symposia that we organized at the 1981 and 1983 meetings of the American Educational Research Association. At those symposia, Tom Shuell and Merl Wittrock acted as discussants. After the chapters were rewritten for this volume and the additional chapters prepared, we asked Shuell and Wittrock to act as discussants once again, this time for the book. Theirs was a most demanding task: to read all the chapters in their respective parts and to provide a concise synthesis. We believe that readers will appreciate their efforts, as their contributions enhance the value of the collections. For ourselves, we are very grateful to them.

Our very special thanks are due to Jim Mackenzie, a philosopher from Monash University, who read the final manuscript and, in the process, produced copious notes and comments. We could not do justice to all of Jim's suggestions, and he must bear no responsibilities for the changes that he advocated but that we failed to make; we wish to place on record our conviction that all the chapters are better in their final form as a result of his work

Not all of the contributions that we originally planned for this volume could be incorporated, notably a chapter by Rosalind Driver from the University of Leeds and Gaalen Erickson from the University of British Columbia and a chapter by Joseph Nussbaum from the Hebrew University of Jerusalem. Our intellectual debt to these and other scholars who for many years have worked outside the mainstreams of educational research to produce the substance of this book cannot be overestimated; we only can hope that their academic and scholarly influences and contributions are adequately evident throughout the volume.

We acknowledge the help of all of those at Academic Press who encouraged and enabled us to bring the idea for this book to fruition. We are indebted to our typists, Kay Smith, Helen Keogh, Dianne Wetzel, and Marilyn Chandler, who were always prepared to adapt their busy schedules to accommodate our often-tight deadline demands.

INTRODUCTION

Leo H. T. West and A. Leon Pines

Two trends in education and psychology have merged to give rise to the exciting new area that is the topic of this book. These are (1) the dramatic shift to cognitive psychology with its interest in the learner-in-the-process-of-learning, and (2) the methodological shift towards qualitative studies. Suddenly, in the late 1970s, and in many countries, studies began that involved in-depth investigations of small numbers of learners about the nature of their knowledge and their learning. Only recently have these studies begun to find their way into the literature, often into journals of the disciplines investigated, rather than in the education or psychology journals. This is not unexpected. These studies are, by their nature, discipline related. Conclusions from an investigation of the problems inherent in learning a concept in physics, for example, are of interest to all physicists who teach physics.

At the same time, there is much in this research that is of vital importance to educators, psychologists, and to most teachers. This book is relevant for all these audiences. Most of the chapters draw on research, rather than report research; they emphasize the contribution of such research to our understanding of the learning process; they provide guidelines for teachers in their quest to have their students achieve meaningful learning.

Throughout the book the emphasis is generally on science learning. It was not our intention to focus on science learning, but rather to focus on the learning of coherent bodies of knowledge. The sciences, especially the physical sciences, consist of well-developed and highly structured bodies of knowledge. This has made them the traditional content area examined by philosophers interested in the nature of knowledge, and, for the same reasons, the prime area of interest for psychologists interested in cognitive learning. The discussions and findings in this volume are not restricted to science learning. They apply to all the areas of curriculum where students are involved in learning a coherent body of organized knowledge.

The title "Cognitive Structure and Conceptual Change" is meant to capture two directions of approach, although like any such distinction it is

COGNITIVE STRUCTURE AND CONCEPTUAL CHANGE   1

necessarily artificial. Part I, which roughly corresponds to the cognitive structure part of the title, is subtitled "Eliciting and representing school-type knowledge." It concerns work that focuses on descriptions of what learners know. The treatment is not primarily methodological, although important methodological approaches are described. The various chapters also raise issues of the nature of knowledge and knowing, the stability of learners' pre-conceptions, methods of changing a learner's conception, and the role of affect and commitment in conceptual change. The chapters are arranged so that the emphasis on methodology decreases from Chapters 2 to 7. One of the immediate results of the type of research represented in Part I was the demonstration that learners' existing conceptions are very resistant to change, that despite extensive instruction and acceptable (even outstanding) performances on school examinations, many students cling tenaciously to their naive notions.

The chapters of Part II, "Stability and Change in Conceptual Understanding," address the question of how to assist and encourage learners to change their minds. As part of this, new ideas about the nature of learning and the role of teaching emerge. As in Part I, the chapters are arranged in a sequence based on changing emphasis. The early chapters have a stronger emphasis on specific examples (based on theoretical models) of instructional approaches aimed at encouraging effective cognitive learning; later chapters place more emphasis on the theoretical models.

The above is a brief overview of the content of the book, but what of its themes? What patterns emerge from the mosaic of ideas? We see the emergence of a theory about the learning that we refer to in the title as "conceptual understanding." We are emphasizing the learning of a coherent body of knowledge, not as discrete concepts, skills, and so forth, but as a related set.

Despite convention, it is appropriate that the discussion that follows is personal and subjective. It is in the same spirit of the chapters in this volume. Conceptual understanding is making one's own sense of knowledge. It involves the learner in constructing ("generating," Wittrock, Chapter 15, would say) his or her own understanding. We present, therefore, our particular "sense making" of the book.

We find it useful (following Vygotsky, 1962) to identify two sources of knowledge in the individual. There is the knowledge that a child acquires from interaction with the environment. We might call this intuitive knowledge, 'gut' knowledge, naive knowledge. It is influenced by language, by culture, by other individuals, and so on. Such knowledge is a person's own sense making of the environment she observes, tempered and manipulated by her interaction with parents, peers, television, and other influences. It primary characteristic is that it constitutes the person's reality. It is something she believes. Another characteristic is that this type of knowledge is acquired in a rather haphazard fashion, over considerable time, and without any particular direction. We do not set out to learn the nature of the earth as a cosmic body, for example. We know that the earth is flat to our eye, yet round from photographs from space. We know about satellites, the shuttle, and a whole gamut of other things. At any time in our lives we have a particular conception of the world which we believe, think of as reality, 'know' is shared by our peers.

The other source of knowledge is formal instruction, disciplined knowledge, school knowledge. It is someone else's interpretation of the world, someone else's reality. Its primary characteristic is authority. It is 'correct'; it is what the book says; what the teacher says. It is approved by a whole bunch of other people who are usually older and more highly regarded than the student. Our learning of this knowledge is goal-directed. That is, we set out to learn, usually through instruction, a particular body of knowledge. We are usually expected to learn it in a certain time period. We are usually expected to demonstrate, most often through tests, what we have learnt about it.

Learning is viewed by the authors of this volume as the process in which learners make their own sense of inputs. Learning always involves the interaction between the learner's present understanding of the world and the knowledge input. Of course, much of what passes for learning is quite different from this. One can read in Chapter 2 by Gilbert, Watts, and Osborne, Chapter 3 by West, Fensham, and Garrard, and Chapter 5 by Champagne, Klopfer, and Gunstone, as well as in the burgeoning alternative frameworks' literature (see, e.g., Driver & Erickson, 1983), that the learning of isolated pieces of knowledge to which the learner has no commitment is commonplace. But we are concerned here with genuine attempts to make sense of school knowledge.

Let us return to our two sources of knowledge and introduce a metaphor that we find useful. Our view of genuine conceptual learning is the integration of knowledge from these two sources; a vine metaphor is helpful in understanding this integration. We imagine two vines representing these different sources of knowledge, the one originating from the learner's intuitive knowledge of the world (which we call the upward-growing vine to emphasize that this is part of the growth of the learner), the other originating from formal instruction (which we call the downward-growing vine to emphasize its imposition on the learner from above). Genuine conceptual learning involves the intertwining of these two vines. The postulated vine metaphor emphasizes that once integration occurs, the sources of particular parts of the intertwined vines are impossible to identify—indeed, at that point, the question of source may be irrelevant. At the point of integration, however, the sources of knowledge are most important.

One can imagine different situations that arise, depending on the nature of our two vines. Initially we identify four extremes.

1. Conflict situation

Both vines are well established but they are in conflict. In this situation, the learner's reality—the ideas that he believes and to which he is committed—is in conflict with the principles being presented, which carry with them the authority of the discipline and the endorsement of the school. Mature learning in this case involves transferring one's commitments from one set of ideas to another. It demands the questioning of one's reality, the abandonment of ideas that have been established over a long period—ideas that one knows are still held by members of one's subculture.

2. Congruent situation

Both vines are well established but not in conflict. In this situation, the student's reality can be integrated with the school knowledge without special problems. There is no reality shock, no need to abandon old commitments. There is simply an extension and an integration of one's reality into a bigger perspective. When the school knowledge is presented, it merely reinforces existing ideas, integrating them into a larger whole, extending one's understanding of the world.

3. Symbolic knowledge situation

This is the situation where there is hardly any upward (intuitive knowledge) vine to interact with the downward (school knowledge) vine. An example of this is the learning of much of freshman, organic chemistry (West et al., Chapter 3). There is little in the learner's intuitive knowledge that is relevant to the learning of, for example, the substitution reactions of benzene and its derivatives. For the student, this is an attempt to acquire pure symbolic knowledge. Even laboratory classes do not provide many elements of reality. It is doubtful that many freshmen chemistry students have seen, felt, or smelled benzene yet they spend considerable time studying its chemistry.

4. Uninstructed situation

This is the case where there is little or no formal school knowledge vine, where all of the learner's knowledge is based on intuitive learning. This is a fascinating area for study, and research cited by Hewson (Chapter 10) demonstrates the powerful influence of cultural metaphors in shaping concept acquisition.

The preceding rather idealized typology can be used to integrate the contributions of the authors in this volume. Acquiring relatively large bodies of complex, inter-related knowledge in school settings probably always begins as a symbolic knowledge situation. Given the authority and the demands of schooling, novices are usually forced to begin by ignoring their own reality. So, if the learner wants to make a genuine effort to make sense of this knowledge (as opposed to rote learning numerous isolated knowledge bits), she will need to concentrate on integrating and differentiating the symbolic knowledge. This is what we see to be the concerns of West et al. (Chapter 3), White (Chapter 4), Pines (Chapter 7), Novak (Chapter 12), and, to some extent, Head and Sutton (Chapter 6) and Champagne, Gunstone, and Klopfer (Chapter 11). Instructional strategies and devices like concept maps (Novak, West et al.), vee maps (Novak), sentence completion, self-concept measures (Head & Sutton, Chapter 6), and a number of others, are designed to encourage and reinforce integration and differentiation of the various parts of school knowledge being presented to the learner. The functional image elicited by the vine metaphor of these procedures is that of increasing the intertwining and consolidating of the various branches in the downward-growing vine, here and there allowing a single runner to push ahead, maybe intertwining with a runner from a different branch, or forming the beginning of a new expanding growth. The general image is of gradual expansion and increasing intertwining. It might be useful to dub this process *conceptual development*.

At the extremes of this developing vine there will occur interactions with the upward-growing vine of the learner's real-world knowledge. Sometimes these will be conflict situations; at other times congruent situations. Gilbert et al.'s (Chapter 2) Interview-About-Instances (I. A. I.) technique and Reif's (Chapter 9) ancillary knowledge strategies seem applicable to the former. If the expanding school knowledge vine meets a runner of the real-world knowledge vine which creates a conflict, what does the learner do? It is not a major reality shock of the kind that might confront a flat earth believer who sees a round world from space, but rather a minor conflict which on its own does not *have* to be resolved. For example, the learner can bypass the conflict by deciding to have two meanings for a word like 'work'—one for school physics, and one for the world of experience. Or, of course, the learner can try to create a relationship between these two meanings, or try to see the physicist's reason for retaining the word 'work' but redefining its meaning. For many learners this would both resolve the conflict and open up avenues for further integration of the two vines. Instructors who prefer the latter, find techniques like the I. A. I. attractive as instructional tools as Gilbert et al.'s workshops for teachers have demonstrated. It might be useful to dub this process *conceptual resolution*.

What happens when the conflict situation is major? When the conflict is not just between one branch of one vine and one branch of the other, but between whole vines? Here we have substantial reality clash. The classic examples of an Aristotelian-like view of mechanics (which is a common intuitive framework) challenged by Newtonian physics, and the Newtonian view, when eventually accepted, challenged again by Einsteinian physics, are both discussed in this volume (Chapters 5, 11, and 13). That the resolution of such conflicts can be painful and therefore difficult to accomplish is well illustrated in studies reported by Champagne et al. (Chapter 11) for instance.

The term *conceptual change* is used by several chapter authors to describe this process (Champagne, Gunstone, & Klopfer, Chapter 11; Strike & Posner, Chapter 13; Johansson, Marton, and Svensson, Chapter 14). The nature of conceptual change is developed at a general theoretical level by Strike and Posner, and at a pedogogical level by Champagne et al. There is considerable congruence between their treatments. To give just one example, Strike and Posner imply that conceptual change strategies will only be successful if substantial understanding of conceptual knowledge has been achieved. Champagne et al.'s conceptual change strategies worked well with university graduates but were unsuccessful with middle school students who had only minimal understanding of the conceptual knowledge.

Let us recapitulate our personal understanding of the themes developed in this volume. The book is concerned with conceptual understanding of the coherent bodies of organized knowledge that form a significant part of school and college curricula. The authors share a constructivist view with respect to such learning. They consider that genuine conceptual learning occurs when learners make their own sense of such knowledge.

There is something of a contradiction here. The curricula of schools and colleges are other people's knowledge, imposed (with the power of authority) on the student. Not surprisingly, some students do not bother to make personal sense of this knowledge but merely play the school 'game' of rote learning and reproducing the curriculum knowledge. But what of those who genuinely attempt to understand school knowledge? How do they make personal sense of it? We have found it useful to distinguish three processes: conceptual development, conceptual resolution, and conceptual change. They provide, for us, alternative structures for interpreting the contributions to this volume.

Conceptual development involves integrating and differentiating the many concepts of a segment of school knowledge without particular regard to one's beliefs and commitments to ideas about the world. It seems that conceptual development can constitute the first step toward conceptual understanding. Techniques for describing an aspect of an individual's conceptual development, instructional strategies aiding conceptual development, and descriptions of the process of conceptual development are found in chapters by West *et al.;* White, Head & Sutton, and Pines, & Novak.

Conceptual resolution concerns the separate concepts and involves resolving differences in meaning between the real-world uses of those terms and their curriculum use. Recent research has demonstrated the extent to which real-world meanings can dominate this interaction and lead to the persistence of major misconceptions. The extent and nature of these misconceptions highlights the importance of instructional strategies to aid conceptual resolution. Chapters concerned with the description of conceptual resolution include those by Gilbert *et al.*, Hewson, and Reif.

Conceptual change involves abandoning one's commitment to one set of conceptual understandings by adopting another irreconcilable set. This abandonment is not always a component in conceptual learning, but when it is, it is a difficult and painful process which requires both a commitment on the part of the learner and special instructional techniques. Detailed discussions of the process and some of the techniques can be found in the chapters by Champagne et al. (5 and 11), Johansson et al. (14), and Strike and Posner (13).

In sum, we hope that this book will provide interested researchers and practitioners with some methods for describing cognitive structures and the means for attempting to effectively bring about conceptual change in their students. In a nascent and burgeoning field like this, many exciting new additions are bound to appear in the framework that we have offered. We feel fortunate to have contributed a vanguard base for future development.

REFERENCES

Driver, R., & Erickson, G. (1983) . Theories-in-action: Some theoretical and empirical issues in the study of students' conceptual frameworks. *Studies in Science Education, 10,* 37–60.
Vygotsky, L. S. (1962). *Thought and language.* E. Hanfmann & G. Vaker (Trans.) Cambridge MA: M.I.T. Press.

ELICITING AND REPRESENTING SCHOOL-TYPE KNOWLEDGE

ELICITING STUDENT VIEWS USING AN INTERVIEW-ABOUT-INSTANCES TECHNIQUE

John K. Gilbert, D. Michael Watts, and Roger J. Osborne

INTRODUCTION

The realization that students bring idiosyncratic meanings for words which are commonly used in science to the science classroom is of long standing. The possible consequences of this for teaching have also been established for some time—Ausubel (1968), for example, refers to them as preconceptions (that are) amazingly tenacious and resistant to extinction, implying that such views are, in some essential way, not only wrong but bad. More recently, however, another interpretation has emerged: Driver and Easley (1978), referring to them as alternative interpretations, observe that "In learning about the physical world, alternative interpretations seem to be the product of pupils' imaginative efforts to explain events and abstract commonalities they see between them" (p. 62). Whatever response teachers decide to make to them, private meanings are a genuine part of children's culture.

If a word's meaning is held before that word is met in formal science classes, we would refer to it as being part of children's science. Across a population, a range of meanings would exist for a particular word. Every child would employ a range of words, the sum of which would constitute a large part of the individual's children's science (S_{ch}). That meaning for a word which was held by a consensus of the scientific community we would describe as being part of scientists' science (S_{sc}). That version of scientist's science which is selected by curriculum planners for inclusion in a syllabus or is enshrined in a textbook could be called curricular science (S_{cr}). The interaction between curricular science and a teacher produces teacher's science (S_t). In classrooms, children's science and teacher's science interact to produce student's science (S_{st}). These representations of knowledge can be articulated (Zylbersztajn, 1983) into the sequence shown in Figure 2.1.

If children's science is a major influence on the curriculum, then attempts

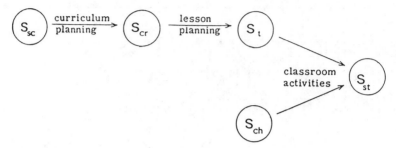

FIGURE 2.1   Transformations of scientific knowledge.

to elucidate its characteristics are called for. Investigators seeing their task as the measurement of the extent to which students have learned the correct— or scientist's science—version of a word meaning will use comparative methods, for example, multiple-choice and open response questions. At the opposite end of the scale, investigators taking an ethnographic, or non-comparative approach, will wish to examine the use of words in real-world settings. The most usual and complex of such settings is the normal school classroom: both Tasker (1980) and Zylbersztajn (1983) are conducting such studies. The simplified classroom, that is, one where two or three students are given lesson tasks to perform in a separate room, has also been used (e.g., by Tiberghein, 1980). The middle ground between these extremes, which is represented by methodologies capable of adaptation to either a comparative or non-comparative mode, seems potentially fruitful. Certainly open response questions can be used in this way (e.g., Viennot, 1979), as can a modified multiple-choice approach (e.g., Helm, 1980). It is into this middle ground that we have introduced the Interview-about-Instances (I.A.I.) technique (Gilbert & Osborne, 1980b; Osborne & Gilbert, 1980).

In outline, the I.A.I. technique consists of tape-recorded dyadic discussions between the researcher and a student, using a deck of cards and focusing on the applications of a single word. A card consists of a line-drawing of a situation which may, or may not, represent an example of the application of the word. Whatever the student's response, reasons are sought. This method elicits a range of responses, as an example will show. Take, as an illustration, the card about cycling from a deck of cards dealing with force. (Figure 2.2)

Typical answers are

"Yes . . . the wheels are still going so that there would be a force from that."   (aged 9)
"It is just putting force on by itself . . . from the force you gave it before."   (aged 11)
"Yes . . . the speed that he [sic] has already got up."   (aged 19, Teachers College student who
                                        passed U. K. "O" Level in Physics at 16)

NO BRAKES
NO PEDALLING
SLOWING DOWN

IS THERE A FORCE ON THE BIKE?

FIGURE 2.2   An example of an I.A.I. card.

"There is a force because of the bike's own mass . . . the mass of the bike has come (to) such a speed that it won't just stop straight away . . . the force is still in there . . . in the bike . . . the force was transferred from the person pedalling and it is now still adherent in the bike . . . the bike still moves forward."

(aged 20, Teachers College student who passed U. K. "A" Level in Physics at 19)

A deck of such cards, perhaps 15 in number, will reveal the breadth and manner of use of the chosen word over a range of situations.

DESIGNING A DECK OF I.A.I. CARDS

The initial design of a deck of I.A.I. cards may be undertaken by following the algorithm presented below (Watts, 1980a, 1980b, 1980c):

Identify the Scientists' Science Meaning of the Word

For force, three component viewpoints can be isolated: "force as pushes and pulls," "resultant force equals mass × acceleration," "forces are interactions of nature." Combined and extended, these produce the conclusion that "a force results from interactions between bodies capable of changing the velocity, size, of shape of the bodies."

Analyze the Scientists' Science View

This implies an exploration of the accepted meanings of body, interaction, and changing velocity. It also involves analyzing the view for its vagueness and ambiguity after the manner of Quine (1960) and Lachenmeyer (1971).

The vagueness of a view is a function of the extent to which the range of object predicates, which form its referential meaning, have been specified; that is, the distance between its connotative meaning and its denotative meaning. For force, the level of vagueness is low: its meaning rests partly on that of interaction which can be specified in one of four distinctive modes: strong, weak, electro-magnetic, and gravitational. The ambiguity of a view depends upon the extent to which it has multiple, equally legitimate meanings. In a physics context, the ambiguity of force is low, but for its societal meaning, the value can be said to be higher. For example, "a person was forced to agree with another," and "a person was forced to do what another person wished."

The analysis also involves clarifying the distinction between basic quantities and derived quantities, the latter being derived from, and defined by, the former. However, the base which is chosen is usually a matter of convenience. This analysis recognises the convenience of defining force in terms of mass and changing velocity.

Identify the Criterial Attributes of the Word in S_{sc}

Criterial attributes are the essential qualities, all of which must be recognized if the word is to be used in a way acceptable to scientists. For force, these involve specifying

1. its magnitude and direction,
2. the body on which it acts,
3. the body that exerts the force,
4. the nature of the force, for example, the type of interaction occurring, and
5. the effects of the force, for example, changes of direction, size, shape.

Identify the Non-criterial Attributes of the Word in S_{sc}

These are statements which are sometimes made about situations or circumstances and which involve the word in some way, for example, for force:

1. the weight of an object can be said to act through its center of gravity,
2. tension is the condition of a body subjected to equal but opposite forces which may lead to an increase in linear dimensions,
3. forces can be measured in Newtons,
4. surface tensions can be accounted for in terms of inter-molecular forces.

The number of these statements is usually great (around 30 for force, at school level), the range being decided essentially by the researcher after examining a sample of appropriate curriculum materials.

Identify Sources of Obvious Linguistic Confusion

These may be grouped into one of three types: antonyms, synonyms, homonyms. Antonyms for force include weakness, incapacity, enfeeblement, tameness. Synonyms for force include strength, power, impetus, violence, intensity, effort, military might, body of people, drive, population, validity. Homonyms for force include fours (quadruples), fauces (a cavity of the mouth) and faucet (a tap). Any of these might conceivably enter into a potential student's answer: cards should be designed with these possibilities in mind.

Identify Sources of Invalid Use of the Scientist's Science Meaning

The first, and probably major, source is that pool which can be called common usage meanings. These are inevitably numerous; examples for force include force of habit, force of law, forced labor, forced march.

A second group is that which may be termed misuse of the scientist's science meaning. Warren (1979) has identified a collection of misuses of force:

1. the supposition of force where none exists, to account for effects noted in a situation;
2. the omission of vector components;
3. an inadequate description of the bodies exerting the forces in the situations in which the interactions are taking place.

If 'examples' of a word may be defined as instances where it could be accurately applied in the S_{sc} meaning, then 'non-examples' may be defined as instances where it cannot be so applied. For force, instances which could be considered non-examples might be those where it is not possible to specify the following:

1. the forces acting,
2. the nature of the forces,
3. the magnitude of the forces, and
4. the direction of the forces,

that is, where one or more of the criterial attributes is not able to be identified.

Producing Cards

The parts in this sequence, which have already been outlined above, might be designated as an exploration of potential sources of alternative conceptions. The actual production of cards is a mixture of art and science. What is required is a mixture of examples and non-examples. Each example would represent all the criterial attributes and some of the non-criterial attributes; for a deck of cards, the intention would be to include most or all of the non-criterial attributes. Each non-example would have one or more of the criterial attributes absent, yet include some of the non-criterial attributes: in a deck of cards the whole spectrum of combinations of criterial attributes would be omitted, and all the non-criterial attributes included.

Therefore, the design of individual cards will depend on the insight of the researcher. *The overriding concern must be to have cards which simultaneously allow* S_{sc}, S_{ch}, *and* S_{st} *to be demonstrated.* Thus, they must be interesting to the interviewees. Ideas can be sought from the illustrations in textbooks, from television programs, and by the simple expedience of asking students and teachers for challenging situations.

Ordering Cards into a Deck

For interviews, particularly with students who are shown to have adopted the S_{sc} view, it has been found that the optimum deck contains fewer than 20 cards; with younger and less knowledgeable interviewees, the activity is appropriately truncated in light of the following sequence:

1. Cards 1–8: examples.
2. Cards 9–12: non-examples. At this point the interviewee is asked for an explanation or definition of the word in his or her own words.
3. Cards 13 and onwards include more difficult instances, likely to be novel to the interviewee but looking rather like textbook physics and including borderline examples.

However, the final identification of cards and their ordering into a sequence is an interactive process forming part of the investigation design.

THE PROCESS OF ELICITATION

After designing the deck of cards, a series of pilot interviews is conducted with about 5–10 students covering the particular age range to be investigated. The purpose of these interviews is to remove simply anomalies in design, wording, and sequencing of the cards. Then a larger trial of about 15–20 interviews is conducted in which fully detailed transcriptions are made. An inspection of these will show whether individual cards, and the sequence adopted, best facilitates the presentation of the *student's own interpretation*. This leads to a final review of the deck, after which the investigation proper is conducted.

The choice of a sample of students with whom to conduct interviews is an initial challenge. The I.A.I. approach may be seen to be a kind of case study. The problem is, what is the case? If the case is an individual student, who will perhaps be interviewed about a number of words, then any articulate individual is satisfactory. If the case is the whole of a naturally occurring group (e.g., a class of students) then no problem is encountered, for all are interviewed. However, if a selection is to be made from a class, then its basis needs careful consideration. As Pines (1980) has pointed out, interview techniques work best with articulate students and these may not be cognitively representative. If the teacher is asked to select a number of "about average" students to be interviewed, then the articulation factor is compounded by the natural desire to please the researcher by providing a satisfactory student: this often means a good one. The use of auxiliary tests to select students (e.g., Piagetian tests, verbal reasoning tests, and so forth) seems equally problematic because the relation between scores on these and children's science is unclear.

Interviews are best conducted in complete privacy, for only then are participants relaxed (and then only if they feel unthreatened). The amount of extraneous noise on a tape-recording will also be reduced to a minimum. The participants sit side-by-side while the interviewer sets the scene for the interview. We have found it useful to say something like: "You will know from your own experience how teachers sometimes use words in classes which do not agree with your understanding of those words. I want to find out how you use a particular word. . . . There are no right answers or wrong answers . . . so that we can get teachers to see your point of view." The interviewer then asks permission to record the discussion, "because I can't concentrate on what you are saying and take notes"; this is almost always acceptable.

The interview takes the form of a discussion. From an initial fairly open-ended question based on the first card, such as "is there a force here?" the interviewer attempts to identify, as accurately as possible, the student's perceptions. The response of the interviewer is non-evaluative, with supplemental questions being asked until the interviewer has fully grasped the student's ideas. Many of the skills needed, such as the reflective quotation, are standard in the area of counselling. As progress is made through the deck of cards, reference back to earlier cards is made either at the student's behest, to change an earlier response, or by the interviewer, if there is a glaring discrepancy between the answers given to two cards. At the end of the sequence, or whenever the interviewer has decided that further progress is unlikely, the student is offered an opportunity to revisit any of the cards. By this time the student is usually still relaxed but a little tired. The interviewer's final questions (although they can also be asked at the beginning) concern the student's prior academic achievements, age, class, previous experience of science teaching, and interest in physics. At no point is the student given any evaluation of success or failure.

THE TRANSCRIPTION OF INTERVIEWS

The transcriber faces a number of challenges in this vital task. First, the discussion is discontinuous, being punctuated by gestures, and faltering and irregular pausing; it shows various qualities of voice, and is delivered with a variety of facial expressions and in many body postures. Second, the sequence of conversation is difficult to follow, being full of inconsequence, confusion, pauses, and contradictions. And third, the tape recorder, whilst ignored for the majority of an interview, can occasionally become the focus of the student's attention. In short, it is inevitable that a transcript becomes a translation prepared by the transcriber.

Whilst the interviewer will have made notes on some of these features, which should be immediately related to a transcription prepared very shortly after an interview, the transcriber will need to use some precise conventions. Of these, the two most important concern the *style* and *notation* of the transcript. In style, we have found it convenient not to punctuate the transcript nor to divide the speech into sentences, and to omit fullstops, commas, and the like. The question mark is also omitted, for the verbal inflective associated with questioning is sometimes used to make statements, the truth of which is uncertain to the speaker. Pauses are typed and marked: noises (e.g., laughter) are included. The transcription is prepared with wide margins (for subsequent annotation) and marked with revolution-counter numbers from the tape recorder.

The transcription is best done in two distinct phases. First, the words and noises are presented verbatim. Then the tape is replayed, entering intonations, pauses, and comments. Irrespective of who transcribes the first part, it must be checked by the interviewer, who must undertake the second part.

The notation that we use is as follows:

1. Square brackets [] enclose information and comments added on the right hand side of the page.
2. Parentheses () enclose interpretations, where unavoidable, after the utterance to which they apply.
3. Transcriber's doubt (*) is shown by asterisks in parentheses with as much of the sound included as possible. The number of asterisks indicate the number of words.

Example 1

```
                                        ┌[utterance number]
045   27←ˉI      what did you have in mind
↑     28 ↗N     its not . . . no its (sh**)
│   [speaker]                  [undecipherable]
[Tape rev-counter number]
```

4. Periods indicate pauses. One period is used for a very short pause, thereafter the number of dots reflect the length of pause.

5. The number of seconds are indicated for long pauses by a numeral preceded and followed by two periods: . . 8 . .

Example 2

053   32   N   its a . er . . . force of some sort . . 5 . . I think.

6. A number of colons included in a word show a prolonged, or drawn out, sound.

Example 3

064   46   I   yes if someone tries . . . . well . . fo::rce you to do it.

7. Stressed syllables—this is done by underlining.

Example 4

066   48   N   oh but no that's different . . . . power is not the same thing at all.

8. A single square bracket [indicated overlapping utterances or where a second speaker cuts into another.

Example 5

```
102   63   I     did you . . . . . ⌈ I mean
      N                          ⌊ yes it was easy to keep doing it
```

Example 6

```
121   76   N     and the force is this   ⌈ way and
      77   I                            ⌊ I see
      78   N     not the way it seems here
```

9. Rising tone / marks a rising inflection, not necessarily a question.
10. Falling tone \ marks a falling inflection.
11. Exclamation mark ! is used at the end of an utterance considered to have exclamatory intention.
12. Umms and ers are included as appropriate. A whistle or a sigh—the sort you might let out at the end of a hard day's digging in the garden (whew!)—these are included as whhhh.
13. Laughter, a snort, a cough, and so forth are included as (laugh), (snort), (cough), and so forth.

THE ANALYSIS OF A TRANSCRIPT

The first task of the editor (usually the interviewer) is to impose a structure on the transcript by identifying and separating the discreet utterances made. We have identified five categories in these units which are concerned with different types of talk and function in the interview. These are A, Personal; B, Task; C, Card; D, Concept; E, Framework. They are elaborated below.

A. Personal

This is all the dialogue that takes place to relate the individuals in the interview. It includes the greetings, introductions, "What courses are you doing?," "What do you want to do when you leave school?" It includes the institutional, social talk that starts and ends the meeting. The function of the talk is the necessary softening of approaches and opening of communication channels. The function of the category is as a collection of dialogue that does not fall into the other boxes. It is useful, too, for providing the overview to the whole interview. Both the personal and task categories are necessary if all the data are to be accounted for.

In network terms it can be abbreviated to[1]

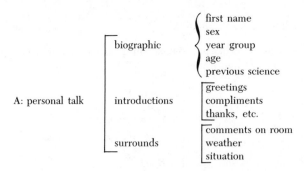

B. Task

In this category, the talk concerns the task: what to do, the confidentiality of the tape, the nature of the survey, the topic of the cards, an so on. This occurs throughout the interview. The function of the talk is to establish the role-play of the task and to cement the "contract" of the interview. The success of the mission, in particular as it relates to the strategies the students adopt, will depend on this part of the session. The function of the category is, again, as a fairly loose framework for small talk.

---

[1]N.B. { indicates "and"; [indicates "or."

C. Card

It is easier to describe the function of this category before describing the type of talk it involves. When a youngster is faced with one of the cards, the concentration fluctuates between what is happening in the card, to what happens generally, and then to the wider scheme of things. We wish to be able to separate local effects and observations from underlying beliefs or "articles of faith." This category we call "card," and it relates together all the talk that we think is a local effect, that is, an immediate response to details or aspects of a particular card. These context-specific responses also have another function. They carry all the evidence for the next category (D: the concepts). They bear all the hallmarks of articles of faith. For example, we quote Jonathan, 14–15 years old. The card concerns a golfer hitting a ball, and the 'card' category responses have been italicized.

```
129   115   J    there'll be quite a few forces acting on that ball
      116   I    yeah what are ⌈ they
      116.5 J               ⌊ they're the thrust of the heating drag of the resistance
                 . . . . yes . . . . there'll be gravity pushing it down and . . . . when it
                 hits the ground there will be momentum keeping it going and it will be
                 leaving far more fric by far more kinetic energy by friction because it's
                 going through the air because the air is much much more er . . . . less
                 dense than the ground its . . . . so what few forces acting on
                 ⌈ there
                 ⌊ tell me about
      117   I    the thrust one, you haven't you didn't mention that one before
                 ⌈ what's the
      118   J    ⌊ thrust is because the push is the actual thing powering it which in
                 fact could be the club making it go so the power is transferred from club
                 to ball and then lost in the ground.
```

The student, Jonathan (J) starts with 115 "there'll be quite a few forces acting on that ball." This is a discrete utterance in that the student seems to have communicated a particular piece of understanding. It is also card specific. It means that he has signalled that the card represents his idea of forces. Secondly, he has singled out the ball, not the golfer. After I 116 "what are they?" he gives ". . . they're the thrust of the heating drag of the resistance . . ." That in itself is another meaning unit. It is also card specific, in that he doesn't refer to this topic again in the interview.

D. The Concept

This category consists of two kinds of talk. It is explicit talk that generalizes *between* cards, for example, in defining types of forces, giving examples of forces in different situations, making statements about a general concept of *force,* and so on. It is also a combination of the implicit hallmarks that indicate a particular conception, in the above example, an implicit statement lies behind Jonathan's utterance 118. According to the Aristotelian view of forces, the golf ball needs a "driving force" to get it through the air. Jonathan's view is basically Aristotelian-like. When asked about thrust (117) he moves quickly from thrust to power, via the word powering. His entire statement is a meaning unit that says that the ball needs a driving force to make it go. This is a part of specific card talk but also heralds the underlying force/motion conception. It reappears in two more cards and so it is something general to the cards.

The meaning unit has at least two functions. It *labels* some aspect of a card or situation; and it also holds a *meaning potential*—it might be a harbinger to a particular alternative conception. This can be best decided if more than one instance occurs to reinforce a particular view.

E. The Framework

Children may have specific answers to specific cards (card talk) and they may have starting points for arguments (concept talk). But they also include a wide range of ideas from outside the cards to substantiate their conceptions. This is framework talk. For example, Jonathan has an extensive framework in that he incorporates the terms "less dense air," "power," "resistance," "thrust," "drag," and so forth, with some semblance of meaning into his explanations. It is this web of ideas—the causes and effects of forces—that form the wider understanding of the pupil.

This category includes two kinds of talk. First, it includes talk that expands the conceptual framework of ideas; and second, it includes all the talk that alludes to projections, speculations, or consequences of the situation depicted in the cards.

In this chapter, we can do no more than provide a taste of the complexities of data analysis. Indeed, one of the major problems in this type of work is to provide a suitable platform for its detailed discussion. The format outlined above for classifying the interview discussion is an attempt to do no more than that.

RESULTS OBTAINED BY THE I.A.I. METHOD

We have identified five distinctive types of understanding of words used in science which are found across a large number of words. A brief sketch of each now follows (Figure 2.3) and includes an explanation of the type, a reduced form of the card concerning the word *force* by which the explana-

(a) Everyday language: A word in science is made sense of by using an everyday interpretation of it.
The man is trying to move the car but the car is not moving.

Is there a force on the car?
Student's view: "Yes, because he is forcing the car." (9, 11, 13)
Physicist's view:

Reaction   Zero

Man's push ←————+————→ Friction (stationary)

Pull of gravity

(b) Self-centred and human-centred viewpoint: Words and situations are considered in terms of human experiences and values.

No brakes
No pedalling
Slowing down

Is there a force on the bike?
Student's view: "No, not really, because he is not pedalling or anything." (9, 11, 13)
Physicist's view: Reaction

↑————→   Friction   ————→
                                      (decelerating)

Pull of gravity

(c) Non-observables don't exist: A physical quantity is not present in a given situation unless the effects of that quantity, or the quantity itself, is observable.

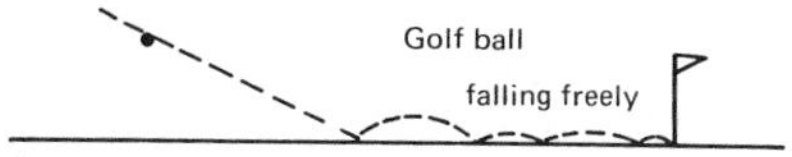

Golf ball

falling freely

Is there a force on the golf ball?
Student's view: "No, it's dropping freely." (7, 9, 11, 13)
Physicist's view:

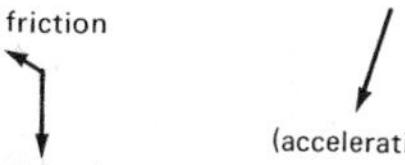

Air friction

↓   ↗
                  (accelerating)

Pull of gravity

(d) Endowing objects with characteristics of humans or animals: Objects are endowed with feeling, will, or purpose. These statements are often not metaphorical.

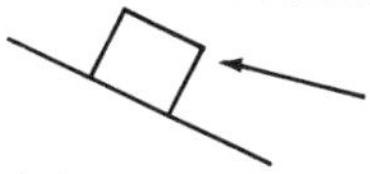

Box   ←——— The box is not moving.

Is there a force on the box?
Student's view: "I suppose there is a bit of force because it has got to force itself to stay up." (7, 9)
Physicist's view: Reaction

Friction ↗   ↖ ↗   Zero (stationary)

Pull of gravity

(e) Endowing objects with an amount of a physical quantity: An object is endowed with a physical quantity which is given an unwarranted physical reality.

Golf ball

Is there a force on the golf ball?
Student's view: "Yes, because the man would be hitting the ball and there would be a force on the ball which would be getting less as it goes up." (9, 11, 13, 15, 19)

Air
Friction ↖

↓   ↗

Pull of gravity   (accelerating)

FIGURE 2.3 Five types of understanding of words identified by I.A.I.'s.

A person throws
a tennis ball
straight up into
the air just a
small way.

The questions are
about the total
force on the ball.

If the ball is on the way up, then the force on the ball is shown by which arrow?

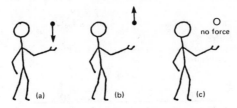

If the ball is just at the top of its flight, then the force on the ball is shown
by which arrow?

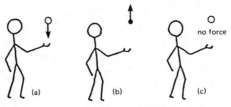

If the ball is on the way down, then the force on the ball is shown by which arrow?

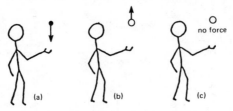

FIGURE 2.4   Multiple-choice questions used to check representativeness of interview data.

tion was elicited, typical quotes from students (their ages are given), and an outline of the scientist's view.

THE REPRESENTATIVENESS OF DATA OBTAINED

In case study work, particularly that concerned with the representation of cognition, it is difficult to draw parallels to the notions of reliability and validity taken from the psychometric tradition of research. Test–retest reliability seems unlikely to be applicable, for the mere process of being interviewed about a word seems, after some reflection, to cause a revision of the understanding held. Parallel-form reliability seems more operable, in that it should be possible to design parallel versions of the same I.A.I. card deck. However, the problem of separating two identical groups of students seems intractable. Face validity seems assured by the group nature of the preparation of card decks. Concurrent validity has not yet been explored; however, work is underway to compare the utility of the I.A.I. technique with the Kelly Repertory Grid technique (Kelly, 1955). Construct validity has been investigated for demonstration-type interviews (e.g., Archenhold, 1980) with respect to Piagetian stages. However, the significance of these results remains unclear while the mechanisms for the development of both word understanding and Piagetian stage promotion are uncertain.

In the meantime, it is only possible to check, using survey techniques, to see if the patterns of student understanding identified by interview means are replicated over large sample sizes. Watts and Zylbersztajn (1981) used a combined multiple-choice/explanation format for such an investigation into forces associated with movement (see Figure 2.4). A summary of some results obtained using this approach in the United Kingdom and New Zealand

TABLE 2.1   Summary of Some Results Obtained in the United Kingdom and New Zealand

Response pattern (see figure 2.4 for meanings of symbols)	Response patterns for student age groups (%)				
	13–14 ($N = 254$)	14–15 ($N = 195$)	15–16 ($N = 174$)	16–17 ($N = 147$)	17–18 ($N = 75$)
bca	46	53	66	53	52
bba	4	1	1	—	—
baa	14	11	9	4	1
aca	9	5	5	14	21
aaa	5	11	6	2	19
acb	11	10	10	7	7
(other)	11	9	3	—	—

is given in Table 2.1. In this case, the combined results confirmed the impressions gained from the analysis of transcript data.

THE FUTURE USE AND DEVELOPMENT OF THE I.A.I. APPROACH

Some attempt is being made in present research to study children's egocentric views about the world through their own scientific conceptions, and in particular to investigate how their views change, and at what points in the child's development such changes occur. Additional research is under way which is aimed at examining

1. the retention of intuitive ideas in children despite formal teaching of those ideas;
2. the language used by students in discussing scientific concepts, and the way in which this language reflects their understanding;
3. the extent to which and the ways in which attitudes and orientation towards science can help in the development of scientific concepts;
4. the effects of different learning styles on the development of concepts;
5. the relationship between concepts, which reflects the alternative framework (Zylbersztajn 1983) developed by the students;
6. the possible existence and influence of sex differences in concept development.

However, we are most concerned that these results can be communicated to teachers. Thus we have developed three forms of a practical workshop (focusing respectively on force, particulate matter, and living) each of which runs for about half a day and which can accommodate primary, secondary, and tertiary teachers simultaneously (Gilbert & Osborne, 1980a). We are also concerned with investigating the implications of these results for teaching (Gilbert, Osborne, & Fensham, 1982). On the development side, we need to supplement the present use of the I.A.I. technique as a research tool by exploring its diagnostic value in the hands of teachers, and its potential as a self-monitoring device for use by students of different ages and at different levels of conceptual development.

REFERENCES

Archenhold, W. F. (1980). An empirical study of the understanding by 16–19 year old students of the concepts of work and potential in physics. In W. F. Archenhold, R. H. Driver, A. Orton, and C. Wood-Robinson (Eds.), *Cognitive development research in science and mathematics* (pp. 228–238). Leeds : Centre for Studies in Science Education.

Ausubel, D. P. (1968). *Educational psychology : A cognitive view*. New York : Holt Rinehart, Winston.

Driver, R., & Easely, J. (1978). Pupils and paradigms: A review of literature related to concept development in adolescent science students. *Studies in Science Education, 5*, 61–84.

Gilbert, J. K., & Osborne, R. J. (1980a). From children's science to scientist's science : A workshop (physics) (chemistry) (biology). Surrey : I. E. T. University of Surrey.

Gilbert, J. K., & Osborne, R. J. (1980b). 'I understand, but I don't get it' : Some problems of learning science. *School Science Review, 61*, 664–73.

Gilbert, J. K., Osborne, R. J., & Fensham, P. J. (1982). Children's science and its conse-quences for teaching. *Science Education, 66*, 623–633.

Helm, H. (1980). Misconceptions in physics amongst South African students. *Physics Educa-tion, 15*, 92–105.

Kelly, G. A. (1955). *The psychology of personal constructs*. New York: Norton.

Lachenmeyer, C. (1971). *The language of sociology*. New York: Columbia University Press.

Osborne, R. J., & Gilbert, J. K. (1980). A technique for exploring students' views of the world. *Physics Education, 15*, 376–379.

Pines, A. L. (1980). Protocols as indications of cognitive structure: A cautionary note. *Journal of Research in Science Teaching, 17*, 361–362.

Quine, W. (1960). *Word and object*. Boston: M. I. T. Press.

Tasker, C. R. (1980). Some aspects of the student's view of doing science. *Research in Science Education, 10*, 19–23.

Tiberghien, A. (1980). Modes and conditions of learning. In Archenhold, W. F., Driver, R. H., Orton, A., Wood-Robinson, C. (Eds.), *Cognitive development research in science and mathematics*. Leeds: Centre for Studies in Science Education.

Viennot, L. (1979). Spontaneous reasoning in elementary dynamics. *European Journal of Sci-ence Education, 1*, 205–221.

Warren, J. W. (1979). *Understanding force*. London: John Murray.

Watts, D. M. (1980a). A transcription format. Unpublished Manuscript, University of Surrey, I. E. T., Surrey.

Watts, D. M. (1980b). Exploration of the concept of 'Force.' Unpublished manuscript, Univer-sity of Surrey, I. E. T., Surrey.

Watts, D. M. (1980c). Suggestions for generating and designing cards for the Interview-about-Instances approach. Unpublished manuscript, University of Surrey, I. E. T., Surrey.

Watts, D. M., & Zylbersztajn, A. (1981). A survey of some children's ideas about force. *Physics Education, 16*, 360–365.

Zylbersztajn, A. (1983). *A conceptual framework for science education : Investigating curricu-lar materials and classroom interactions in secondary school physics*. Unpublished doc-toral dissertation, University of Surrey, Surrey.

3

DESCRIBING THE COGNITIVE STRUCTURES OF LEARNERS FOLLOWING INSTRUCTION IN CHEMISTRY

Leo H. T. West, Peter J. Fensham, and Janice E. Garrard

INTRODUCTION

What does it mean to know a complex subject like undergraduate chemistry? What is it that students must internalize from, say, a freshman science course, in order to be considered knowledgeable? Certainly less than the professor, but less in what way? Such questions are integral to any attempt to describe the cognitive structure that learners develop as part of a regular sequence of instruction. We have undertaken a series of studies in various areas of chemistry with the general aim of eliciting and representing the cognitive structures, both of individual learners and of groups of learners (West, Fensham, and Garrard, 1982). The studies all concerned relatively short sequences of instruction (3–4 weeks of the normal curriculum) and covered three topic areas in freshman chemistry (amino acids and peptides, one-component phase diagrams, and radiochemistry) and one topic in year 11 high school chemistry (mixtures and compounds). In this chapter we draw on that work, concentrating primarily on answers to questions of the kind raised above, but also providing some examples of the cognitive structure representations achieved and a brief outline of the investigative techniques that evolved.

LEARNING, KNOWLEDGE, AND COGNITIVE STRUCTURE

Any attempt to describe cognitive structure is influenced by, among other things, the investigator's theories, whether explicit or implicit, about learning and knowledge and the nature of cognitive structure. These theories are also influenced by the investigations. It is impossible for us, now, to separate our pre- and post-investigation theories. What is presented here is a post-investigation description of our theories, prejudices, speculations, etc., about learning, knowledge, and the nature of cognitive structure that developed with, and were essential for, the investigations.

LEARNING AND KNOWLEDGE

The nature of knowledge has interested philosophers for over 400 years and it is certainly not easy to separate the epistemological from psychological aspects of it. Nonetheless, we have found it useful to distinguish between public knowledge and private understandings, relating them (approximately) with knowledge and the result of learning, respectively. Science exists as public knowledge, in text books and in scientific papers. When individuals read (or are told) this public knowledge, they interpret and internalize it in their own way. We have all experienced this. We read a paper, and understand it. Then we read it again a while later and find we understood it much more. Our "understandings" between these two readings is different. So too are the "understanding" of other people who read the paper. In reading a paper (or in learning any public knowledge via any learning mode) we may miss some bits of the information or we may put the bits together in different ways from that presented or implied. Further, when we learn any bit of information, we relate it to our own previous knowledge and experience, so that the understanding of each of those bits of knowledge, the meanings that we give to them, are idiosyncratic to some extent. Thus, our private understanding is not fixed; it is not the same as that of other people; it is not the same as the public knowledge.

We should now turn this discussion on its head. Public knowledge is derived from the private understandings of individuals. Public knowledge exists because there is a substantial overlap between the private understandings of different individuals. Sutton (1981) explains it this way:

> Though a metal does not mean exactly the same thing to you as to me, there are many common features, upon which we can form an agreed definition—a metal is conducting, formable, tending to be electropositive, etc. However, such public knowledge represents only a part (the denotative part) of the meaning that particular persons carry, and is to some extent an abstraction, not existing in any individual. (p. 3)

Public knowledge is defined and definable. In chemistry, at least, two important types can be identified: propositions and algorithms. A parrot can be taught to utter propositions. So can computers, although computers more commonly perform algorithms. Calculators, for example, usually add, multiply, calculate square roots, and so forth, although there are some 'calculators' on the market that can be asked questions and can give answers (i.e., they can state propositions).

These things are knowledge but they are not learning (see, for example, Popper, 1974, pp. 147–148). Learning is giving personal meaning to public knowledge. It is developing one's own understanding of that public knowledge. Computers do not learn—they can state propositions and they can do calculations (or other algorithms), but they do not give meaning to these acts. (There are some exceptions to this generalization, especially in the field of artificial intelligence. For example, Quillian's TLC (Teachable Language Comprehender) was designed to infer meanings from simple sentences (see Quillian, 1969).

The computer metaphor can be further exploited. Given below is a segment of an interview transcript.

Men are all alike
IN WHAT WAY
They're always bugging us about something or other
CAN YOU THINK OF A SPECIFIC EXAMPLE
Well, my boyfriend made me come here
YOUR BOYFRIEND MADE YOU COME HERE
He says I'm depressed much of the time
I AM SORRY TO HEAR YOU ARE DEPRESSED
It's true. I am unhappy
DO YOU THINK COMING HERE WILL HELP YOU NOT TO BE UNHAPPY
I need some help, that much seems certain   Weizenbaum (1966)

The above conversation is not between two people. It is between a young lady and a computer program called ELIZA. ELIZA's conversation is given in capitals. The program was developed by Joseph Weizenbaum of Massachusetts Institute of Technology.

It would be easy to think that this program is listening to the young lady, analyzing what she is saying and following that up with relevant questions. It actually is not. It operates on a few simple grammatical rules, avoiding the need to comprehend any of the interviewee's answers.

We use the example here to illustrate what happens in communications. ELIZA has no private understandings at all. It has no concepts. The young lady, however, does. Yet in the conversation she infers meaning into what ELIZA says, meaning that was never present in ELIZA's store of information. For example, look at the young lady's response to "I am sorry to hear you are depressed." She replies, "it's true, I am unhappy." ELIZA, of course, has no such meaning for "depression." Indeed, ELIZA has no meaning at all for "depression." It simply used one of its grammatical rules to produce the sentence. But the young lady does have a private understanding for being depressed. From her answer we know that she associates it with being unhappy, and she uses this private understanding to give meaning to ELIZA's statement "I'm sorry to hear you are depressed."

Teaching is much more complex than this, but it contains the same elements. There is a syllabus that contains the "public knowledge" to be taught (and in science this is usually propositional or algorithmic knowledge). The teacher has private understandings of this public knowledge, which he aims to share with his students. One cannot expect the teacher to share all of this private understanding, however. Suppose one part of the syllabus is the definition of a mixture. The teacher has a wide experience of mixtures—if he is a chemist he has tried to separate them, seen their effect as impurities in melting points and in IR spectra, studied their effect on phase properties (phase diagrams), and on crystal structure; and many, many more. All of these bits of knowledge and experience give meaning to the teachers' understanding of this definition being presented to the students. They are there "in his head." He may not even think of them, but they influence his private understanding of the definiton just the same. He cannot expect to communicate all of those to his pupils. He will probably attempt to give his pupils some experiences, and he might raise their consciousness of some of their own experiences and knowledge of mixtures. But even in the teaching of a single definition, the teacher cannot hope to communicate the richness of his own private understanding of that definition to the students. Most students' private understandings (insofar as they are derived from the teacher) will necessarily be poorer than the teacher's.

Definitions such as that of a mixture are never learned in isolation. Mixtures will be taught as part of a series of lessons that teach other definitions and skills. These various bits of knowledge are not learned in isolation. They are related, and the students are expected to relate them together. Again, the teacher's private understanding contains more than what is to be taught. If a teacher were to list all the other things that come to mind related to mixtures, the list would be much greater than the list of things that are part of any syllabus topic containing the concept "mixtures." In the teacher's mind, there are a large number of other related knowledge bits. Note that we have used the terms "other" and "related." There is a knowledge structure, if you like, in which the knowledge bits are all interrelated in some way (the teacher's cognitive structure). Not only will the teacher communicate only some of these knowledge bits to the pupils, but the teacher will also communicate only some of the relationships. What the teacher will try to share with these pupils is a slice through personal cognitive structure or private understandings.

Let us summarize what we have said to this point. There is public knowledge that exists in science textbooks and papers. A syllabus topic specifies some parts of this that are to be taught. In science, these will include definitional statements (or implied definitions like "mixture") which are propositions, and skill statements (e.g., calculate the pH of a weak acid solution) which are algorithms. There may be also laboratory skills and some other ways of specifying the public knowledge. The teacher has a private understanding of this part of science, which he or she attempts to share with the pupils. The pupils will develop their own private understanding which will be poorer than the teacher's, partly because they will not internalize all of the bits of information, partly because the inter-relational links will be less extensive, and partly because there will be fewer other experiences, knowledge, and other skills to add meaning to each new bit of information. Any individual pupil's private understanding will, of course, also have some idiosyncratic features.

The Nature of Cognitive Structure

Cognitive structure is extensive and n-dimensional. Any attempt at description can only hope to illuminate part of this whole. The n-dimensionality inevitably leads to a trade-off between the extent and the detail of the description. If the major interest is in detail, then extent must be sacrificed—only a small portion of cognitive structure can be described. If extent is important then detail must be reduced. Our balance of these two lies somewhere between the associative network descriptions of people like Lindsay and Norman (1977) and the concept maps of people like Novak (1980) and Rowell (1978). We have limited our extent to a small segment of a student's regular learning, and we have not searched for the detail that the Lindsay and Norman representations require. To some extent we overcome some of this trade-off using a notion that we have called "node compression." This notion will be described below.

We are also very aware that to focus on one segment of cognitive structure distorts the description. We are taking a "slice" (and a non-planar one) through a learner's n-dimensional cognitive structure. This slice is then presented in isolation. This may produce many distortions. For example, certain knowledge may be stored hierarchically under more general ideas (or subsumers if we use Ausubel's language), but this structure may be excluded in the particular slice that we take.

Cognitive structure has two components—the knowledge bits it contains, and how that knowledge is organized. Knowledge bits can be quite different in size and nature. Since we wanted to represent cognitive structure as a diagram with nodes representing knowledge bits joined to other nodes, we needed to make some decisions about what should be used as nodes. In

Lindsay, Norman, and Rumelhart (LNR) representations, the proposition "A is a B" is represented with two nodes A and B connected by a line labelled "isa." Such a system, derived from each proposition, is not practical in teaching components of 6–8 lectures which will contain hundreds of propositions. We needed to find some way of reducing the number of nodes (and as a consequence, increasing the amount of implied knowledge associated with a node), and at the same time retaining the flexibility to represent the extent of that implied knowledge when that was desirable. This problem of bit size was resolved through the node compression notion mentioned previously. Only a brief description of the term is given here.

When we receive input through our senses, we have to infer a great deal from that input. Consider the sentence "Bob Hawke is the Prime Minister of Australia," and imagine it is spoken to an intelligent machine which had relatively little information stored in its memory. Such a machine might need to ask, "Am I to infer that Bob Hawke is a name that is associated with a person?" or "Is Prime Minister a title which may refer to a position?" If this sentence was spoken to an American, it is possible that he or she would use his or her knowledge of the nature of the position of President of the United States to give meaning to the term Prime Minister. This would indeed help give meaning to the sentence, but it might also distort the intended meaning as the two positions are not equivalent. In fact, the listener needs to infer a great deal and this ability to infer depends upon information stored in the listener's storage memory. We can take this point further by considering two part sentences:

"The window in the bathroom. . . ."
"The stove in the bathroom. . . ."

In order to understand the first of these, the listener does not need to bring to mind everything he knows about windows and bathrooms. The listener will have no difficulty making sense of the sentence, and will assume that he or she and the speaker have the same meaning for "window" and "bathroom." However, in the second sentence there is more difficulty with the inference. What kind of bathroom has a stove? The listener might make an historical inference—there was a time when bathrooms contained a wood-fired heater that might be described as a stove. So the listener might infer that this is what the speaker had in mind. Or he or she might know that the speaker is a bachelor living in cheap "digs" and so infer that this is a curious room that serves as both kitchen and bathroom, or he or she may ask for clarification. What is important is that the listener has only begun to investigate this deeper knowledge of bathrooms and stoves when such an investigation was called for—and then only to the extent that it was necessary. Thus, although a whole complex cognitive structure relating to bathrooms and stoves and to the speaker may exist in the listener's head, he or she has used only the simplest compression of that structure that was sufficient for the inferential task.

This leads to the "node compression" notion. The term 'node' describes a knowledge bit of indefinite size in memory. When we concentrate on a specific node in order to use it for interpreting an input, for example, we have "compressed" under it all of the complex cognitive structure that we have linked with that bit in our memory. That "compressed" knowledge will only be called upon when it is needed for inference—and only to the extent that it is needed. Thus, we can have available all of the richness that we have stored as part of our knowledge of "bathrooms," but we do not need to consciously "bring to mind" any of that richness that is not needed.

It is important to recognize that the compressed node, no matter what or how much it has compressed under it, is the label that is used in communication—and that the listener infers meaning about the communication from *his* cognitive structure compressed under the node and *not* that of the speaker. Though our labels, or compressed nodes, are shared, our meanings for them are idiosyncratic.

This whole notion of node compression is derived from the "frame" idea of Winograd (1975), among others. It is really saying nothing more than that although we may have an elaborate sub-network associated with specific nodes in our cognitive structure, we are able to compress that network into a single node when this is convenient. However, this idea is very powerful in dealing with the question of node size, and in resolving some difficult aspects of representation.

With regard to the nature of the knowledge bits, most of the representations to date have used *semantic networks* that are primarily *propositional.* This is despite the quite considerable evidence that images are part of memory (e.g., Paivio, 1971), and the practice of teachers in teaching algorithms as processes to be remembered. Gagné and White (1978) proposed four types of memory structures (as they called them), namely, propositions, images, episodes, and intellectual skills. Lindsay and Norman (1977) have also incorporated events (episodes) into their semantic networks.

We have used three types of knowledge bits (based on Gagné and White) as nodes—propositions, algorithms (or skills)[1] and, in one study, images. We have also used examples, a special type of proposition, as separate nodes.

Before proceeding to describe some of the outcomes of the research, there

---

[1]We use the term *algorithm* when we refer to public knowledge and skill (or intellectual skill) when we refer to private understanding. This is not just a matter of semantics. Books cannot have skills. They can only outline the steps of an algorithm that can be used to perform a particular task. A person who can perform that task, possesses that skill, whether he follows the book algorithm or not.

is one other distinction that we have made. In science courses, there is frequent discussion of "concepts," such as "amino acid," "radioactivity," and "phase." Words which we might describe as concept labels occur in the propositions that make up the public knowledge of a subject. The meaning of the concept for any person, however, is part of his or her private understanding. Thus, a person has a particular understanding of the meaning of a concept label like "amino acid," which will usually include the public knowledge definition (although at times students learn a different definition), but this understanding is more than that proposition. The student's understanding includes all the other knowledge that he relates to that proposition in order to infer meaning. For an extreme example of the dependence of meaning on the private understanding of a sentence we quote the following story of a child after the first day at school who complained to his mother,

"They never gave me a present."
"Present what present?"
"They said they'd give me a present."
"Well now, I'm sure they didn't."
"They did! They said 'You're Laruie Lee, aren't you? Well you just sit there for the present.' I sat there all day but I never got it. I ain't going back there again!" (Donaldson, reported in Gilbert and Osborne, 1980).

Here, Laurie has internalized the sentence "you . . . sit there for the present," but has given a very different meaning to it than that intended by the speaker.

As humans, we develop cognitive structures all the time. Each time we observe something we have the potential to add that to some part or parts of our cognitive structure. In doing so we enhance the meaning of the observation and of other things we have already stored in cognitive structure. However, in our research, we were interested in intended learning. We wanted to investigate the cognitive structures developed as a result of the teaching of specific public knowledge. We assume that the intention of teaching is to develop certain new knowledge bits in the learners' cognitive structures, that are inter-related in certain ways. That is, we assume that there is an intended knowledge structure that is to be developed as part of each learner's cognitive structure. In practice, different learners will internalize this intended knowledge structure with different degrees of completeness, different degrees of accord with the intended meanings, and with different nature and degree of relationship with other aspects of their cognitive structure.

We are hinting here at the obvious. Cognitive structure is highly idiosyncratic. It is indeed possible to explore an individual's cognitive structure using very unstructured approaches, but the results of such a description will be very individual. Since one of our aims was to explore the possibility of summarizing individuals' cognitive structures on some general dimensions, it was necessary to reduce the degree of idiosyncracy. Intended learning is an obvious way of doing this. For most teaching sequences, there is a clearly definable set of new public knowledge that all learners are intended to learn (where we interpret that to mean "incorporate into their cognitive structures"). This intended cognitive structure part was used as the focus in our investigations. It provided a framework for the description. This does not deny the possibility of investigating other knowledge—and indeed this also formed an important part of the investigations (although this aspect is not explored in this chapter).

We return to an idea that we mentioned earlier. What we are investigating is a slice (and a non-planar one) through the learner's cognitive structure. The intended cognitive structure is really a slice through some expert's cognitive structure that has been translated into a part of a course. The curriculum of this part-course is limited when compared to the expert's cognitive structure. There is much in the expert's cognitive structure that is related to the elements of learning in the course that is not taught, nor even articulated (compressed under the nodes to use our term). There is much other knowledge to which the elements of learning relate, that are also not part of the course. Some of these will be part of other courses; some will be part of the expert's experience and general knowledge; and some will be things he or she has discovered.

The learner in this course is not going to be presented with all of this richness. He is going to be presented with the "slice." The underpinning of his nodes will be poorer than that of the expert. His general knowledge probably will be also restricted. It may be that a student merely internalizes the slice (or part of it) and makes no attempt to relate it to other aspects of cognitive structure. It is more likely that some degree of linkage occurs.

At this point, our two perspectives have converged. In investigating the cognitive structure of students that has resulted from a segment of teaching, the starting point is the public knowledge. We can define that from such things as the syllabus, the lectures, the examination papers, and hand-outs to students. This will yield the knowledge bits, but rarely the intended structure. (On some rare occasions, one finds "concepts maps" as part of the syllabus, but usually the syllabus and other sources only yield a list of knowledge bits with some imposed groupings of those bits). The public knowledge derived from these sources provides the framework to investigate the learner's private understandings. We can investigate, for example, if the learner has internalized each of the set of propositions and algorithms (and in what form), how the learner has related them together, and what has been compressed under each of these new "nodes" by the learner (if anything).

Let us repeat an earlier claim. The meaning of a concept for any person is part of his or her private understanding. Yet different people use the same concept labels. Hence, public knowledge propositions that contain concept labels may seem to be precise (consider, for example, the scientific definitions that are used so often in scientific courses), while the meaning that an individual infers from that proposition depends upon the individual's private understandings of the concepts. We were referring, then, to public knowledge as the public knowledge in a discipline. The same relationship exists, however, between an individual's concepts (the private understanding) and the person's discourse—which are the propositions that he or she has made public. The listener knows exactly what these propositions are - and he knows what he or she (the listener) understands by the words in the proposition, but he or she cannot be certain about what the speaker understands by the proposition. This distinction is more important when two individuals are involved, as is the case when a researcher is trying to describe a learner's cognitive structure. The learner has certain private understandings. A response to a question by the learner produces a proposition. The proposition is not the totality of his or her private understanding. The listener has private understandings for the words in the proposition. This private understanding is richer than the proposition that has been made public, *and* it is different from the private understanding of the individual who made the proposition public.

For this reason, we consider it inappropriate to use a concept label as a node in a representation of cognitive structure. A concept label like "force" is imprecise in that it does not tell us what the student understands by that concept label. If we use a proposition generated by the student as the node then at least the node is precise. We can then explore, if we wish, the learner's deeper private understanding of the proposition.

Thus, our nodes are usually in propositional form. When we do use a concept label as a short hand notation, we intend it to represent a specific proposition produced by the student. This does not imply that we conclude that knowledge is stored in propositional form, only that in our representation of cognitive structure we have chosen to use learner-generated propositions where others have tended to use concept labels. It also does not imply that we consider that knowledge is only propositional, for we have included algorithms and, sometimes, images as part of our representation.

IMPLICATIONS FOR ELICITING AND REPRESENTING COGNITIVE STRUCTURES

We can now set out the implications of our theoretical framework for describing the cognitive structures of the students in a segment of a course.

The starting point is the course itself. We have extracted different types of "public knowledge" from the syllabus, lectures, examinations, and handouts to students.

These are

1. Propositional statements;
2. Algorithms (in Science, these are often called "problems");
3. Examples (a special type of proposition that we have found it useful to isolate, e.g., glycine is an example of an amino acid);
4. Images (visual representations presented as part of the course).

For any lecture or any other form of discourse, there are many propositions. The node compression notion enabled us to concentrate only on a small number of propositions at any time—depending on the specificity we wanted to achieve. Thus, if we wanted to describe a three lecture series, we would aim to extract the 10–20 main propositions. Listed under each of these propositions we could place a series of other propositions of lesser importance. Whether we used the latter as a starting point to investigate a student's cognitive structure depended on the degree of detail we wanted to achieve. The same was not generally true of algorithms and examples—there were generally only a small number of these in any extended discourse (for example, a three lecture series).

We decided to use these propositions as the primary tools in our exploration of cognitive structure as the stimuli for eliciting the components of the learner's cognitive structure and as the nodes in our representation of that structure. In the latter case, we decided to use a propositional skeleton as the underlying representational structure with algorithms and examples linked to it at appropriate points. Each propositional node was used as a stimulus to explore the depth of a learner's private understandings.

COGNITIVE STRUCTURE REPRESENTATION

Before presenting a real example in all its complexity, we present, in idealized form, the manner we used to represent an individual's cognitive structure based on the propositional skeleton. This description leaves aside for the moment the questions of how this information was elicited.

Suppose that the propositional skeleton obtained is that represented in Figure 3.1(a), where the boxes represent learned propositions and the letters represent the perceived relationships between them. Algorithms and examples that the learner has stored are added to this skeleton where linkages to specific propositions have been made. This is represented in Figure 3.1(b).

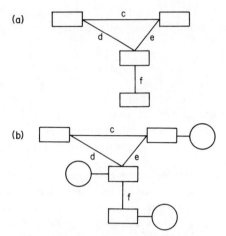

FIGURE 3.1 (a)   Style of propositional skeleton used for representation. Boxes represent
stated propositions, letters represent stated
relationships. (b) System used to add skills
and examples to propositional skeleton.

Each of the propositions may be known as a single piece of verbal knowledge or it may be understood in considerable depth. An additional representation was to list the propositions generated when each of the "boxed"
propositions was explored in depth.

The result of the whole process is a series of representations, each providing different aspects of specificity and/or emphasising different aspects of the
cognitive structure. In most cases these were:

- A propositional skeleton in which the nodes are the propositions generated by the student in response to the free definition questions and the
  connecting lines are labelled with relational terms. (An example is given
  in Figure 3.2).
- A propositional skeleton containing the information above, with learned
  algorithms and examples added. In this form the student generated
  propositions are abbreviated to concept labels. (An example is given in
  Figure 3.3).
- A segment of the propositional skeleton containing the student-generated propositional nodes and a set of propositions produced by the
  student when questioned in depth about the propositional node. (An
  example is given in Figure 3.4).

These representations formed the data base for further reduction, especially for the production of group summaries. Representations of the kind

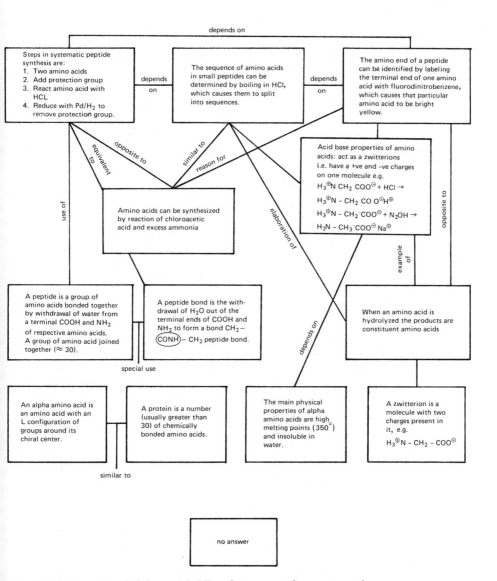

FIGURE 3.2   Propositional skeleton with full student-generated propositions shown.

used in Figures 3.2 and 3.3 and 3.4 can be combined into a single diagram. We have found that this single diagram contains so much information that most people find it very difficult to understand. The set of three diagrams is more useful.

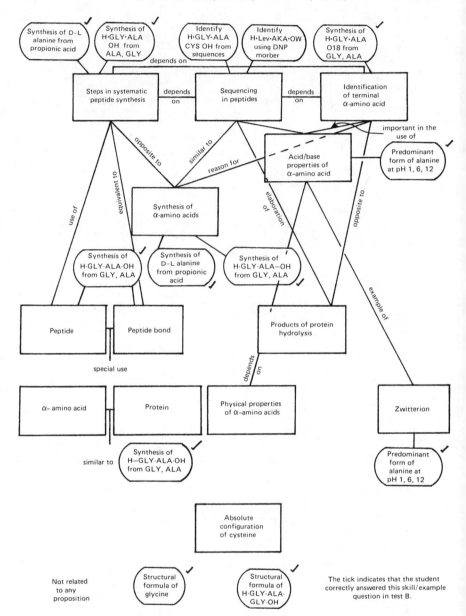

FIGURE 3.3   Some propositional skeleton with skills and examples added (in circles)

> The main physical properties of α-amino acids are high melting point crystalline solids that are, with the exception of a few (e.g., leucine, arginine, etc.) quite insoluble in water. Are amphoteric.

> α-amino acids are found in nature — ones found in protein (L amino acids)

They are crystalline solids, probably white. Amines and carboxylic acids usually exist as low M.P. solids or liquids, usually liquids.

Amino acids are different because there is a combination of a basic group with an acid group to make a crystalline structure which is more stable than a carboxylic acid or an amine on its own.

I don't know in what form the amino acid exists in the solid state.

The amino acids are dipolar. In acid solution there is an extra H on NH_2.

In basic solution there is an H missing from COOH.

In neutral solution there will be both NH_3^+ and COO^-. There might be resonance between NH_3^+ and COO^- within one amino acid molecule.

(Student did not understand that under most conditions amino acids exist as the zwitterion).

I didn't know the answer to this, so I made a guess.

I don't know what the α means.

The general structure of amino acids is

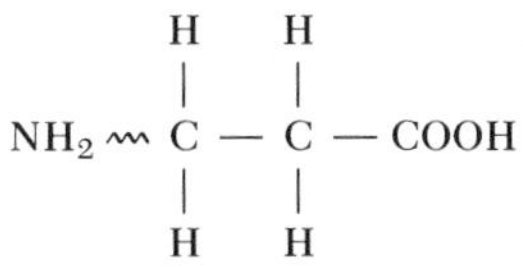

$$NH_2 \sim C - C - COOH$$

with H substituents on each C.

They have two types of structure, L and D, except for glycine. The R group can be methyl, there are some with sulphur, SH.

FIGURE 3.4   One segment of the same propositional skeleton showing depth of knowledge compressed under the nodes.

ELICITING THE DATA

We intend only to give a very brief description of the data-gathering techniques. The propositional statements are obtained using free definition-type questions (e.g., what is an amino acid?), and the skills as typical science problems. The relational link between skills/examples/images and the propositions are obtained using a simple matching task, in which the student was asked to state which (if any) of the skills questions was closely related to each of the free definition questions. The methods of obtaining the propositional skeleton and the depth of knowledge for each propositional node are summarized below. It should be noted that the whole procedure requires about one hour of group testing and one hour of individual interviews.

Obtaining the Propositional Skeleton

The structure of the propositional skeleton is based on inter-relationships that the student perceives between propositions. After attempting a variety of techniques ranging from a card sort procedure (similar to CONSAT used by Champagne, Klopfer, DeSena, & Squires, 1981) through procedures based on multidimensional scaling, we settled on a simple but effective technique. The students were asked to rate all pair relationships on a 0 to 3 scale (which we later collapsed to a 0, 1 scale, representing no significant and significant relationship respectively). This matrix was then tranformed to an hierarchical representation using a simple algorithm. The main components of this algorithm are first to count the number of pair-wise links for each proposition, then to place them on a diagram arranging those with the greatest number of links at the top of the page, those with the next at a level below the top, and so on. Connecting lines are then drawn between the propositional nodes to produce the skeleton. It is usually possible to move nodes horizontally and so minimize crossovers by the connecting lines.

In a follow-up interview, the student is presented with the diagram and asked to describe the nature of the relational links represented by lines on the diagram. The words the student chose to describe the relationships between the proposition are then written along the lines. Figure 3.2 shows a typical example.

Depth of Knowledge for Each Propositional Node

As part of a follow-up interview, the student's free-definition answer sheet was returned to him or her. The student's answers (which bear a one-to-one correspondence to the propositional nodes) were used as the stimulus for further questioning. The format was "can you tell me more about x?" or "You answered Y, can you tell me more?"

Each response was followed up to explore the student's meaning of substantive words. New concepts arising from the students' explanations were explored, while generally staying within the curriculum.

During the interview, the researchers had the set of cards with each main proposition written at the top and all of the associated propositions which were used in the lectures listed underneath. If the student "dried up" without mentioning all of the associated propositions, he or she was asked about it directly with a question like "Can you tell me something about . . . ?"

The transcripts of these interviews were tranformed into independent student discourse. Figure 3.4 shows a typical result.

SUMMARIZING THE COGNITIVE STRUCTURE REPRESENTATIONS FOR GROUPS OF LEARNERS

In some senses, a group summary of cognitive structures is a contradiction in terms. If cognitive structure is idiosyncratic, then what sense can be made of a group summary? In our attempts to represent cognitive structure, we have used the intended public knowledge as a focus, and so have concentrated on those parts of cognitive structure that should be shared among the learners who are part of the course. In this case, there seems to be value in attempting to find some group indexes. These will be useful to teachers who want to obtain feedback about the in-depth learning of their classes. They will also be useful to researchers wanting, for various reasons, to obtain measures of cognitive structure as outcome measures.

With these needs in mind, we have explored ways of converting the cognitive structure descriptions into quantitative indexes. The descriptions, once developed, inform about various aspects of cognitive structure, although, of course, not all possible aspects. Attempts were made to transfer these into numerical indexes. The indexes are necessarily crude, but they still appear to be valuable.

Dimensions on Which Information is Available

White (1980) has provided a valuable contribution in attempting to define various dimensions of cognitive structure and then to seek ways of obtaining measures of those dimensions. We have not set out to use his dimensions, but we acknowledge our debt to his influence. Indeed, we have approached the question from the other direction. We have attempted to describe cognitive structure first and then inspected the description for dimensions that might be extracted.

Apart from two nearly conventional measures (propositional knowledge and skill knowledge), we have extracted five quantitative indexes of dimensions of cognitive structure: integration of propositional knowledge, differentiation of propositional knowledge, differentiation of skills/examples knowledge, articulated propositional relatedness, and depth of propositional knowledge.

Integration and differentiation are two important aspects of cognitive structure. Integration describes the degree to which ideas are inter-related. Since our description is based on a propositional skeleton, integration of propositional knowledge is a dimension of interest. One index would be a simple count of the number of propositional links as a proportion of the total number of possible links. In this case, neither of the extreme scores, 0 or 1, would be educationally desirable scores, although this is the nature of integration rather than of the scales. For example, if a person had every one of a set of propositions inter-related in all possible pair-wise combinations, then the individual would have very poor differentiation (but excellent integration).

However, it would be preferable if the produced index number conveyed some meaning of itself. That requires specifying a certain number of desirable links. One source of this would be the public knowledge. We could put the public knowledge propositions together as an "ideal" propositional skeleton. This skeleton could then be used to provide an ideal number of links which could be used as the denominator in a ratio index. We decided to use such a measure. Integration of propositional knowledge is a ratio of the number of propositional links in the student's skeleton to the number of propositional links in the ideal skeleton. The ideal skeleton was constructed by the researchers after attending all lectures, inspecting all hand out materials, and so forth. It was checked for validity by the lecturers.

Differentiation implies that more specific ideas are subsumed under more general ideas. Since we have used a propositional skeleton and have added skills and examples to this skeleton, we distinguish between differentiation of propositional knowledge and differentiation of skills/examples. For propositional differentiation, we are assuming an hierarchical organization, that is, we are assuming that specific propositions are subsumed under more general propositions that themselves are subsumed under more general propositions. To give an index to propositional differentiation, we need to classification of the generality of the propositions. We use the ideal skeleton to provide that order. We then extract the order of generality from the student's propositional skeleton—those propositions at the highest levels (and therefore linked to more other propositions) are given the highest ranks. The index is then the correlation between the two ranks for each proposition (and is therefore equivalent to the Spearman rank order correlation coefficient).

In the case of skills and examples, there is generally a one-to-one correspondence between a particular skill or example and a particular proposition. A well-differentiated cognitive structure would have each skill and example related to its particular proposition. A less-differentiated structure might have the skill or example related to an appropriate but more general (or more specific) proposition. A poorly differentiated structure might have skills or examples linked to inappropriate propositions. We have used a ratio of $C + 0.5A - I$ to the number of skills and examples (i.e., the total number of correct one-to-one relationships) where C is the number of skills and examples linked to its particular proposition, A is the number related to an appropriate proposition, and I the number related to inappropriate propositions. The use of 0.5 is entirely arbitrary. Negative scores in the numerator were given the value of zero.

The integration and differentiation indexes pick up the existence of relationships. In addition, our maps contain information on the articulated relationships between propositions. Often students have two propositions related, but find it difficult to articulate the nature of that relationship. To provide an index of this, we have scored their articulated relationships into another index. We use a simple count of valid relational words, in which we have subjectively judged the validity of articulated relational words as a ratio of the number of links produced on the student's map. Thus, it is the proportion of perceived links that the student can validly articulate.

The last dimension, depth of propositional knowledge, is a reflection of variations of the extent of underpinning of the curriculum propositions. Some students knew the propositions well but nothing else. Others had deep understanding of their meaning. As an index, we have used the average number of correct, relevant (as judged from the public knowledge) propositions (per main proposition) generated by the student in the interview.

SOME RESULTS OF THE QUANTIFICATION OF COGNITIVE STRUCTURE DIMENSIONS

These indexes of cognitive structure have been extracted in two studies to give some preliminary indications of their form. The first of these studies used only nine students. In the second, involving 22 students, only the group testing was used, so there are no Articulated Propositional Related-

TABLE 3.1 Intercorrelations between the Indices

	Skill knowledge	Integ. P.K.	Diff. P.K.	Differ. S/E	Articulated P.R.
Study with a freshman chemistry students ($n = 9$)					
Propositional Knowledge	.54	−.32	.15	−.09	.61
Skill Knowledge		−.09	−.42	.45	.36
Integrated Propositional Knowledge			−.65	.07	−.07
Differential Propositional Knowledge				−.40	.15
Articulated Propositional Knowledge					−.17
Study with eleventh grade high school ($n = 22$)					
I.P.K.			−.31	−.24	−.18
D.P.K.				−.16	.31
D.S.E.					−.11

ness or Depth of Propositional Knowledge indexes. The inter-correlations between the indexes for both studies are given in Table 3.1.

None of the above correlation coefficients are statistically significant. Because of the small numbers, especially in the first data set, the statistical power is low and there is a good chance that we would make a Type II error if we concluded that there is no relationship between any of these indexes. Of course, we would not want to argue that the dimensions should be orthogonal. It seems reasonable, however, to conclude that there is a degree of uniqueness between the indexes.

The dimension scores derived here have face validity but the data collected for them are insufficient to properly measure their characteristics. At the moment, they must be considered as useful indexes that have potential to be used as dimension scores. It is important to note that only two of the indexes depend on the interview: depth of propositional knowledge and articulated propositional relatedness. The rest can be collected using a one hour group testing, followed by about one hour of analysis per subject. This makes the indexes very practical as measures to use in an experimental study.

REFERENCES

Champagne, A. B., Klopfer, L. E., DeSena, A. T., & Squires, D. A. (1981). Structural representation of students knowledge before and after science instruction. *Journal of Research in Science Teaching, 18,* 97–111.

Gagné, R. M., & White, R. T. (1978). Memory structures and learning outcomes. *Review of Educational Research, 48,* 187–222.

Gilbert, J. K., & Osborne, R. J. (1980). Identifying science students' concepts: The interview-about-instances approach. In W. F. Archenhold, R. H. Driver, A. Orton, & C. Wood-Robinson (Eds.), *Cognitive development research in science and mathematics.* Leeds: University of Leeds.

Lindsay, P. H., & Norman, D. A. (1977). *Human information processing: An introduction to psychology.* New York: Academic Press.

Novak, J. D. (1980). Methodological issues in investigating meaningful learning. In W. F. Archenhold, R. H. Driver, A. Orton, & C. Wood-Robinson (Eds.), *Cognitive development research in science and mathematics.* Leeds: University of Leeds.

Paivio, A. (1971). *Imagery and verbal process.* New York: Holt, Rinehart and Winston.

Popper, K. Autobiography. In P. A. Schilpp (Ed.) *The philosophy of Karl Popper.* La Salle, IL: Open Court, 1974.

Quillian, M. R. (1969). The teachable language comprehender. *Communications of the Association of Computing Machinery, 12,* 459–476.

Rowell, R. M. (1978). *An approach to the evaluation of audio-tutorially instructed children's concepts of selected phenomena.* Unpublished doctoral dissertation, Cornell University, New York.

Sutton, C. (1981, September). *Public knowledge and private understandings.* Paper presented at the Science Education Conference at Pembroke College, Oxford, U.K.

Weizenbaum, J. (1966). ELIZA—a computer program for the study of natural language communication between man and machine. *Communications of the Association of Computing Machine, 9*, 36–45.

West, L. H. T., Fensham, P. J., and Garrard, J. E. (1982). *Final Report: Describing the cognitive structures of undergraduate chemistry students.* Unpublished report presented to the Education Research and Development Committee.

White, R. T. (1980, March). *Converting memory protocols to scores on several dimensions.* Paper presented at the meeting of the Australian Association for Research in Education, Sydney, Australia.

Winograd, T. (1975). Frame representations and the declarative–procedural controversy. In D. G. Bobrow & A. M. Collins (Eds.), *Representation and understanding: Studies in cognitive science.* New York: Academic Press.

INTERVIEW PROTOCOLS AND DIMENSIONS OF COGNITIVE STRUCTURE

Richard T. White

THE NEED FOR DATA REDUCTION

At one time tests used in schools were simple, direct assessments of whether students could recall facts they had been taught or could perform skills in which they had been drilled. Then, quite sensibly, tests began to tap secondary aspects of knowledge: that is, whether students could transfer their knowledge and apply it in solving problems. This shift for a while tended to make people overlook the function of memory of facts and skills as a mediator between instruction and complex performance. The cognitive psychology movement has re-awakened interest in memory, and there is a surge in the attention that is being given to the sorts of knowledge people have and how they store it—in a word, to their cognitive structure.

While a useful construct, cognitive structure is also an ill-defined one. Its definition as the knowledge someone possesses and the manner in which it is arranged raises a number of pertinent questions: In terms of what units or elements is the knowledge to be described? What is meant by arrangement of knowledge? These questions lead to others: What varieties of elements of knowledge are there? What dimensions are necessary for a full description of the arrangement? Researchers' answers to these questions determine how they set out to measure cognitive structure.

Early work (Deese, 1962; Johnson, 1967; Shavelson, 1972) used word associations to probe cognitive structure. Different techniques have been developed since, along with a variety of ways of describing cognitive structures. The most subtle, fine-grained techniques use interviews in one form or another. Examples are found in the work of Pines (1977), Osborne and Gilbert (1979), and Champagne, Klopfer, and Anderson (1980). These interview techniques promise to give us great insights into how people store and recall knowledge and use it in thinking. They provide so much information, however, that there is a danger of drowning in a sea of uninterpretable data. A single one of Pines' interviews, for instance, is so rich in information that it can keep an investigator occupied for weeks, and two of them produce enough data for a doctoral dissertation. For some purposes an investigator may not be able to afford such complexity and subtlety. For instance, cognitive structure may be a dependent or a mediating variable in a sizable experiment, in which a researcher might be interested in comparing the effects on cognitive structure of levels of an independent variable such as a curriculum or teaching method, or might wish to study the relation between cognitive structure and performance on particular tests. For any of these purposes, the protocols obtained in interviews are unwieldy, and will have to be reduced to sets of scores.

The derivation of scores from interview protocols must be done without destroying too much of the sensitivity that the investigator has been at such pains to secure. This requires that a number of well-chosen dimensions must be defined, so that a reasonable description of a protocol can be obtained by converting it to a set of scores, one for each dimension. The number of dimensions needs careful judgment: too few, and too much sensitivity is lost; too many, and again there is danger of drowning in data.

ELEMENTS OF COGNITIVE STRUCTURE

The dimensions that one chooses or defines are influenced by the model of cognitive structure one has in mind. If, for instance, the model represents cognitive structure as fluid and dynamic, there would be a place for a dimension of stability or consistency; in a model that integrated affective components with cognitive, there might be a dimension of commitment to, or confidence in, an area of knowledge. The dimensions that I define below, purely as an example, are based on a restricted model of cognitive structure developed by Gagné and myself (Gagné & White, 1978). The Gagné and White model is a static one, in which there are four sorts of element: propositions, intellectual skills, images, and episodes.

Propositions are representations in memory of facts or beliefs. They are the unit of many memory models, such as those of Ausubel (1968), Anderson and Bower (1973), and Kintsch (1972). Their popularity in models of cognitive structure is easily explained: they are the basis for writing and speaking, our most obvious and important methods of communicating with each other; they are a conveniently-sized unit; their possession is readily tested; and it is easy to devise instruction to give people blocks of propositions sensibly collected together. However, networks that consist of propositions alone can be omitting large sections of relevant knowledge, and for many purposes a more differentiated set of elements may be useful.

Gagné's (1968) distinction between intellectual skills and propositions is the same as that made by Ryle (1949) between knowing that and knowing how, or by Greeno (1973) between propositional and algorithmic knowledge. The essential feature of the distinction is that where propositions are single facts, intellectual skills are rules which direct behaviour so that people can perform whole classes of tasks.

Images are mental pictures. In recent years there has been debate concerning the existence of separate storages in the brain for images and propositions. Paivio (1971) described the operation of imagery in terms of a separate store, but Pylyshyn (1973) argued that a single mechanism for storage of words and pictures is sufficient. Investigators are accumulating evidence which some interpret as confirming the presence of separate stores (Andre & Sola, 1976; Bacharach, Carr, & Mehner, 1976; Kosslyn, 1976; Marschark & Paivio, 1977) and others as denying it (Baggett, 1975). Educators may avoid the present confusion by regarding the debate as one over the fineness of memory unit with which we should be concerned and by accepting that the appropriate level is a matter for individual preference and depends on the purpose for which units of cognitive structure are required. Pylyshyn's assertion of an undifferentiated store may well be correct at the fine level of basic neural structures, but in education it could be more profitable to work at a coarser level which does involve different types. Because the ability to form mental pictures is universal, and because real pictures have long been seen as a powerful mode of communication, images should be included among the elements of knowledge when we are considering instruction and performance.

Tulving's (1972) distinction between episodic and semantic memory has attracted about as much attention as that between images and propositions, but much less controversy. Episodes, the recollections of event, come back as pictures and words, which implies that there is no separate store for semantic and episodic memories. Tulving emphasised that he was not proposing such a separation: "I will refer to both kinds of memory as two stores, or as two systems, but I do this primarily for the convenience of communication, rather than as an expression of any profound belief about structural or functional separation of the two" (1972, p. 384). Episodes should be important in education, because knowledge must be based on experience. Even the most abstruse concepts are given meaning by their relation with real objects, which in turn are understood through personal contact.

Although a dynamic model of cognitive structure may be a more accurate representation of memory, no one seems to have developed a useful one yet. Among static models, the Gagné and White one has an advantage in that its elements are the right size for education. They are units which can be acquired in a brief interchange, unlike the larger elements called concepts which Shavelson (1972) and Novak (1977) use in their representations of cognitive structure, and are meaningful wholes in themselves, unlike the finer modes and relations into which Anderson and Bower (1973) analyze propositions.

PROPOSED DIMENSIONS OF COGNITIVE STRUCTURE

To commence the process of debate about dimensions, I propose a set of nine dimensions based on the Gagné and White model.

One of the most obvious dimensions of cognitive structure is *extent*. Some people know a lot, others little. A more subtle property of knowledge is its *precision*. An example might clarify the meaning of this dimension. Consider a word such as "choreography." People's knowledge of this term could be at several levels of precision; some might have never seen it before; others might recognize it but be unable to do anything with it; some might be able to think with it to some extent, by knowing that "It has something to do with ballet": others might be able to use it correctly; while those with the most precise knowledge might not only be able to use it but also be able to explain its meaning to someone else. Precision applies to single words, propositions, or skills, or to whole bodies of knowledge. Coarse units such as concepts are relatively blind to variations in precision. Two people who associate "force" with "energy" may differ greatly in the precision of their knowledge.

Internal consistency and *accord with reality or generally accepted truth* are related dimensions. As well as being interested in how much a person knows and how precisely he or she can formulate it, we could want to know whether all parts of knowledge are compatible. This may be particularly important for people at the extremes of extent of knowledge of a topic, the tyro and the expert. When someone begins to learn a topic, the new knowledge may conflict with old at points, and it could take some time, as part of the process of learning, for the contradictions to become apparent and to be resolved. The relevance of this to teaching is profound. For experts, contradictions in their knowledge may appear as signals of fundamental errors in their models of reality. While at first these contradictions may be no more than sources of uneasiness, when they become specific they can generate creative advances in the sum of human knowledge.

Much the same point can be made about accord with reality. Bodies of knowledge can be large, precise, and internally consistent, yet mistaken. Discrepancies between knowledge and reality may be again most obvious in the cases of those who know a little and those who know a lot. Given certain purposes, it may be useful to describe someone's cognitive structure in terms of either or both of these dimensions, of internal and external consistency.

Another dimension is *variety of types of element*. Some people are known to possess much "book learning" about a topic, which is another way of saying they have a large proportion of verbal knowledge and little in the way of episodes or skills. Or the imbalance could take a different form; in art, for instance, one's knowledge might consist of many images and episodes of visits to galleries, yet lack any propositions that contain information about the paintings or any intellectual skills such as being able to recognize paintings of a particular style or school. For some topics imbalance might not matter or could even be inherent in the subject, but often the desirable form of cognitive structure will be, as well as of large extent, precision, and consistency, one of a good mixture of types of element. A geographer, for instance, probably needs facts about countries, images of land forms, skills of translating contour maps, many concepts, and recollections of visits to particular places.

As well as the dimension of variety of types of element, there is the dimension of *variety of topics*. Often the purpose of measuring cognitive structure will make this dimension irrelevant, as when one is interested only in knowledge of a delimited topic, but given other purposes it can assume importance. One might want, for instance, to distinguish between people who are specialized in knowledge of a small number of fields and generalists who have some knowledge of many topics. This dimension could be important in comparisons of school systems or curricula, though it does not appear ever to have been assessed.

A dimension which cannot be separated from the specifics of what is known is the form of organization of cognitive structure, or its *shape*. If we think of knowledge as a network of elements, of whatever types, we can conceive of networks having different shapes and degrees of interlinking. For instance, someone might know the following four propositions:

1. Columbus was born in Italy.
2. Columbus thought he could sail westwards to China.
3. China was an important source of spices.
4. Spices were needed to disguise the flavour of bad meat.

The first two are linked by the common term, Columbus, propositions 2 and 3 by China, and 3 and 4 by spices, so the shape of this knowledge is a chain (Fig. 4.1a). But if the propositions

A. Marco Polo had brought spices back from China.
B. Marco Polo was Italian.

are added, the shape becomes more compact and there is greater interlinking (Fig. 4.1b).

In the example above, the shape was changed by adding propositions. It is possible also to imagine two people who know the same things but associate

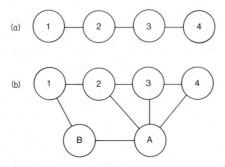

FIGURE 4.1   Effect on shape of adding propositions.

them in different patterns. Where one person associates a certain episode
with a skill, for instance, the other might not. The sudden association of an
episode with a skill or fact is the quite common sensation of perceiving that
some past event is an instance of an abstract principle.

It may be that, to be fully useful, the shape dimension will have to be
refined into several more precise measures. Chains and nets differ in the
number of associations per element, with a chain of n elements having a total
of $n-1$ connections and nets having a greater number ranging up to $\frac{1}{2}n(n-1)$.
Thus, shape could be represented for some purposes by an index called
association density, the average number of associations per element.

Another aspect of linking, which is related to the dimension of variety of
topics, is the proportion of elements in the chain or net which are internal, in
the sense of obviously being parts of the subject matter, and the proportion
which are external, or inessential parts which illustrate the topic rather than
form a vital part of it. External links may be important, even though the
topic is a coherent whole without them, because they relate one topic with
another, so making possible creative leaps, and because they tie abstract
bodies of knowledge to experiences of the everyday world. Mayer and
Greeno (1972) have shown that such links affect understanding. Thus, the
dimension *ratio of internal to external associations* is likely to be important
when considering understanding.

The final dimension proposed here is *availability* of knowledge. Two people may know the same things, but differ in the ease with which they recall
relevant elements at need. The source of such a difference is an absorbing
realm for research, and if explained may lead to dramatic improvements in
human performance. Hunt (1976) has made considerable progress in this
field. In the meantime, speed of recall can be measured without knowledge
of why it differs and can reflect a crucial property of someone's knowledge.

These nine constructs, and the practical measures of them, are not necessarily independent of each other. In fact, it is most unlikely that they would be. Availability, for instance, may well be related to extent or precision or shape, and the description of ratio of internal to external associations presents it as an aspect of shape. Dimensions need not be orthogonal to be useful. The intent here is to propose a number of constructs which may be useful in describing cognitive structure. So, for the present, inter-relatedness of dimensions is not a matter for concern.

THE USEFULNESS OF DIMENSIONS

This beginning set of dimensions, and any other that succeeds it, should meet three tests: practicality, robustness, and creativity.

For a set of dimensions to be practical, it must be possible to convert interview protocols to scores on them. This is not as simple as it sounds: it is all too easy to propose a dimension which cannot be scored. The set described above has been tested for practicality (White & Gunstone, 1980). Interviews, lasting about an hour, were conducted with 28 science graduates on two topics: electric current and eucalypts. The protocols were readily converted to scores on the dimensions of extent, precision, accord with reality, variety of types, and ratio of internal to external connections. In these interviews no consideration was given to assessing variety of topics or availability, though it was not difficult to see that availability could have been assessed by timing responses. Few instances of inconsistent knowledge were observed, though if they had appeared, a simple procedure for producing a score on that dimension was available. The only dimension which was difficult to score was shape, which needs further consideration before the whole set can be regarded as completely meeting the test of practicality. This difficulty with shape is interesting, as shape is one of the terms frequently used in discussions of cognitive structure. It illustrates the danger of the wide use of ill-defined constructs.

A further aspect of practicality is whether the scores mean anything. The procedures used to convert the protocols to scores could work quite well as algorithms without the scores having any relation to underlying characteristics of the people interviewed. Support for the meaningfulness of the proposed dimensions and the scoring procedures is provided by the moderate to very high correlations observed by White and Gunstone (1980) between scores for each dimension on the two topics of electric current and eucalypts. These were chosen as relatively unrelated topics, yet the interviewees' extent scores correlated 0.84, their ratio of internal to external connections 0.47, and the other dimensions in between. This suggests that the dimensions do reflect properties of people's cognitive structures.

The second test that a set of dimensions has to meet is that of robustness against variations in circumstances. The set of dimensions useful in a given circumstance is only one of four interacting factors. The other three are the purpose for which the description of cognitive structure is required, the psychological model used by the investigator and the units of memory which it prescribes, and the method of investigation. There is no simple determining path between these four factors. While purpose does determine the others to an extent, it in turn is affected by what is possible, which is a function of the methods that are available, and they are determined by models and dimensions, though changes in these are brought about by new purposes and methods. It would be remarkable if a set of dimensions was found which was universally applicable; on the other hand, there is little value in a set so idiosyncratic that it can be used only in one specialized circumstance. For ease of communication, we need dimensions which are useful across a range of variations in the other three factors. It is not yet known whether the set I have proposed meets this test. It is, after all, based on the model and units of memory postulated by Gagné and myself (Gagné & White, 1978).

The third test is whether the set enables people to invent new methods for probing cognitive structure, methods which may provide purer measures of single dimensions than do present techniques, which generally provide information about several dimensions at once. This has not yet been tried with my set, but it is a potentially fruitful activity which I hope will attract some attention soon.

I advocate consideration of this issue of dimensions, because there is a danger that the siren call of methods of investigating cognitive structure will lead to much pragmatic collection of data, each study fascinating in itself, but each so singular that there is no correspondence between one researcher's work and another's. Dimensions may enable the separate studies to be linked into a coherent whole, with great benefit to our understanding of learning.

REFERENCES

Anderson, J. R., & Bower, G. H. (1973). *Human associative memory.* Washington, DC: V.H. Winston.

Andre, T., & Sola, J. (1976). Imagery, verbatim and paraphrased questions, and retention of meaningful sentences. *Journal of Educational Psychology, 68,* 661–669.

Ausubel, D. P. (1968). *Educational psychology: A cognitive view.* New York: Holt, Rinehart & Winston.

Bacharach, V. R., Carr, T. H., & Mehner, D. S. (1976). Interactive and independent contributions of verbal descriptions to children's picture memory. *Journal of Experimental Child Psychology, 22,* 492–498.

Baggett, P. (1975). Memory for explicit and implicit information in picture stories. *Journal of Verbal Learning and Verbal Behavior, 14*, 538–548.

Champagne, A. B., Klopfer, L. E., & Anderson, J. H. (1980). Factors influencing the learning of classical mechanics. *American Journal of Physics, 48*, 1074–1079.

Deese, J. (1962). On the structure of associative meaning. *Psychological Review, 69*, 161–175.

Gagné, R. M. (1968). Learning hierarchies. *Educational Psychologist, 6*, 1–9.

Gagné, R. M., & White, R. T. (1978). Memory structures and learning outcomes. *Review of Educational Research, 48*, 187–222.

Greeno, J. G. (1973). The structure of memory and the process of solving problems. In R. L. Solso (Ed.), *Contemporary issues in cognitive psychology: The Loyola symposium*. New York: Wiley.

Hunt, E. (1976). Varieties of cognitive power. In L. B. Resnick, *The nature of intelligence*. Hillsdale, NJ: Erlbaum.

Johnson, P. E. (1967). Some psychological aspects of subject-matter structure. *Journal of Educational Psychology, 58*, 75–83.

Kintsch, W. (1972). Notes on the structure of semantic memory. In E. Tulving & W. Donaldson (Eds.), *Organization of memory*. New York: Academic Press.

Kosslyn, S. M. (1976). Can imagery be distinguished from other forms of internal representation? Evidence from studies of information retrieval times. *Memory and Cognition, 4*, 291–297.

Marschark, M., & Paivio, A. (1977). Integrative processing of concrete and abstract sentences. *Journal of Verbal Learning and Verbal Behavior, 16*, 217–231.

Mayer, R. E., & Greeno, J. G. (1972). Structural differences between learning outcomes produced by different instructional methods. *Journal of Educational Psychology, 63*, 165–173.

Novak, J. D. (1977). *A theory of education*. Ithaca, NY: Cornell University Press.

Osborne, R. J., & Gilbert, J. K. (1979, February). *An approach to student understanding of basic concepts in science*. University of Surrey Institute for Educational Technology.

Paivio, A. (1971). *Imagery and verbal processes*. New York: Holt, Rinehart & Winston.

Pines, A. L. (1977). *Scientific concept learning in children: The effect of prior knowledge on resulting cognitive structure subsequent to A-T instruction*. Unpublished doctoral dissertation, Cornell University.

Pylshyn, Z. W. (1973). What the mind's eye tells the mind's brain. *Psychological Bulletin, 80*, 1–24.

Ryle, G. (1949). *The concept of mind*. London: Hutchinson.

Shavelson, R. J. (1972). Some aspects of the correspondence between content structure and cognitive structure in physics instruction. *Journal of Educational Psychology, 63*, 225–234.

Tulving, E. (1972). Episodic and semantic memory. In E. W. Tulving & W. Donaldson (Eds.), *Organization of memory*. New York: Academic Press.

White, R. T. & Gunstone, R. F. (1980). Converting memory protocols to scores on several dimensions. *Australian Association for Research in Education Annual Conference papers*, 486–493.

5

INSTRUCTIONAL CONSEQUENCES OF STUDENTS' KNOWLEDGE ABOUT PHYSICAL PHENOMENA

Audrey B. Champagne, Richard F. Gunstone, and Leopold E. Klopfer

INTRODUCTION

There is general agreement among physics instructors and students that mechanics is difficult to teach and to learn (Kolody, 1977). Students have difficulty comprehending classical mechanics, and physics instructors often express disappointment with the outcome of their efforts to instruct students in classical mechanics. This instructional problem has been discussed at length in the literature of physics education where various underlying causal factors contributing to the problem have been suggested (Gerson & Primrose, 1977; Halley & Eaton, 1975; Hudson & McIntire, 1977).

In addition to identifying possible reasons for the apparent opacity of physics, and especially mechanics, science educators have attempted various resolutions of the difficulty. Variables that may contribute to students' success in learning physics that have been investigated include (1) mathematical skills, (2) general level of cognitive development, and (3) specific cognitive processes (e.g., Arons, 1976; Renner, Grant, & Sutherland, 1978). The basic strategy used by science educators to investigate these variables is to identify a single student characteristic (level of cognitive development as defined by Piagetian theory, for example) and demonstrate that the characteristic is correlated with success in physics. Then, typically, instruction is modified to take into account student inadequacies with respect to this characteristic, and studies are conducted to demonstrate that student achievement improves. However, this research strategy has produced limited results, as Mallinson (1977), Peterson (1979), and other commentators have pointed out.

A factor contributing to students' difficulties in learning mechanics, in the view of some physicists, philosophers of physics, and, more recently, science educators and cognitive psychologists, is that students come to introductory physics courses with firmly embedded theories of how and why objects move. These naive theories have some characteristics in common with an Aristotelian view of the world.[1] Dijksterhuis (1969) notes that historically a great effort was necessary to outgrow the Aristotelian view of motion and that

> to this day every student of elementary physics has to struggle with the same errors and misconceptions which then had to be overcome, and on a reduced scale, in the teaching of this branch of knowledge in schools, history repeats itself every year. The reason is obvious: Aristotle merely formulated the most commonplace experiences in the matter of motion as universal scientific propositions, whereas classical mechanics, with its principle of inertia and its proportionality of force and acceleration, makes assertions which not only are never confirmed by everyday experience, but whose direct experimental verification is fundamentally impossible . . . (p. 30)

These and other writings by physicists, physics educators, and philosophers of science describe a situation in which certain features of students' pre-instructional knowledge of the subject matter domain is in clear conflict with the canonical view of the subject matter domain which students will be required to learn. This suggests that the problems students experience when learning mechanics derive, in part, from the disparity between their experience-based knowledge about the motion of objects and the facts, principles, and theories of Newtonian mechanics.

In this chapter, we propose an approach to the instructional problem posed by mechanics which takes into account information about the motion of objects that students bring to instruction. We review research that provides descriptions of the facts, principles, and theories of motion held by students who have experienced no formal mechanics instruction, along with research that demonstrates how this knowledge affects students' interpretations of instruction. Results from research contrasting the physics knowledge of expert physicists and novices (individuals who have completed one or two college level physics courses) are used to formulate cognitive objectives for introductory mechanics instruction. Finally, we propose several instructional strategies to facilitate the students' attainment of the objectives. We hypothesize that these strategies will measurably alter the extent, accuracy,

---

[1]Aristotle considered rest to be the natural state of objects. In the absence of any cause, an object does not move; conversely, when an object is moving, its motion must have been caused, usually by a force. Aristotle also argued that the *speed* of an object is directly proportional to the force acting on it, and inversely proportional to the resistance of the medium through which the object is moving. In Newtonian physics, it is stated that an object will continue in its existing state (either ar rest or moving with constant speed in a straight line) unless it is acted on by a *net* force. The *acceleration* of the object is directly proportional to this net force, and inversely proportional to the mass of the object.

and structural organization of the students' knowledge, thus bringing it into a state closer to that of expert physicists.

As we noted, previous approaches focused attention on the generalized thinking and mathematical skills the student brings to instruction. The approach we propose emphasizes the significance of the domain-specific declarative knowledge—facts (for example, the acceleration due to gravity at sea level is 9.8 m/s^2), concepts (acceleration), propositions (acceleration is proportional to net force) and theories (motion occurs only when a force is exerted[2])—that the student brings to instruction.

Two important features of declarative knowledge are elements and structural organization. Elements or structural units of declarative knowledge include concepts, propositions, and schemata. The characteristics of these constructs are not well-specified. Concepts and the relations among them are often represented in node-link diagrams (Anderson, 1983). Two concepts and the relationship between them comprise a proposition. Schemata are larger structural units and are used when representing declarative knowledge about classes of things or events. A playground slide, a stairway, a wedge, and a hill are objects. A child sliding down the slide, a girl walking up the stairway, the wedge splitting a piece of wood, and an elephant sliding down the hill are physical events. A physicist, when thinking as a physicist, conceives of the class of things in this example as inclined planes and the class of physical events as objects moving along inclined planes. This knowledge is represented as schemata—a schema for inclined planes and a schema for objects moving along inclined planes. Schemata have slots. For a specific instance, the slots are filled by information unique to that instance. In the instance of a red metal playground slide, the plane slot is filled by the information red and metal, and the inclined slot is filled by the information that the angle between the plane and the horizontal is about 45 degrees.

The constructs of concepts, propositions, and schemata are used in this paper to describe declarative knowledge about the motion of objects typically used by uninstructed students and by expert and novice physicists.

THE UNINSTRUCTED STUDENTS' EXPERIENCE-DERIVED KNOWLEDGE

Results from empirical studies provide detailed descriptions of uninstructed students' declarative knowledge about the motion of objects. The findings from this research consistently demonstrate that prior to formal mechanics instruction most young people (and uninstructed adults) have a

---

[2]This is an example of an experience-based theory observed in most untutored children and adults. The theory stands in direct contradiction to formal physical theory.

conception of motion that is more Aristotelian than Newtonian (Champagne, Klopfer, & Gunstone, 1982; Champagne, Klopfer, Solomon, & Cahn, 1980; Driver, 1973; Driver & Easley, 1978; Leboutet-Barrell, 1976; Singer & Benassi, 1981). Our conclusion that the pre-instructional concepts are Aristotelian is based on data collected using several demonstrate, observe, and explain (DOE) tasks.

The DOE tasks have three parts: (1) some simple physical apparatus and a manipulation of it are described; (2) the students are asked (a) to predict the outcome of the demonstration and (b) to report the information they used to generate the prediction; (3) the demonstration is then done and the students are asked to describe their observations and to discuss any conflicts between their predictions and their observations.

Our contention that students' conceptions influence their understanding and remembrance of science texts and lectures, their observations, and their interpretations of their observations is also based on observations made during and after administration of the DOE tasks. In one DOE task used in our investigations, students are asked to compare the times for two objects to fall equal distances. The objects are the same shape and size but differ in mass. This task has been administered to several groups of academically talented adolescents (seventh and eighth graders) and to beginning college physics students.

An example from the students' protocols illustrates the interactions between the students' conception of the speeds of falling objects and their interpretation of information obtained from text or lecture. A significant number of the college and middle-school students predict that the heavier object will fall faster and justify the prediction by stating that Galileo proved the general principle that heavier objects fall faster. To support their assertion, students describe an experiment that they recall having learned that Galileo performed—dropping a feather and a gold coin. The students note that Galileo reported that the coin falls faster, thus arguing that Galileo proved that heavier objects fall faster. We infer this to mean that the students have been exposed to some of Galileo's work, either by reading or hearing about it. However, if we compare the generally accepted account of Galileo's argument with the students' interpretation, we note significant differences. Galileo used a thought experiment to support his assertion that when air resistance is controlled for, the heavier object and the lighter object will fall at the same rate. Galileo's complete argument is that if a feather and a coin are dropped in air, the coin will fall faster; if, however, the objects are dropped in a vacuum, the effects of the upward push of the air will no longer be present and both objects will be observed to fall at the same rate. It is our observation that students who predict that heavier objects fall faster remember the part of Galileo's experiment that is consistent with their conception of falling objects and forget the part of the experiment that is inconsistent with their conception.

There is also evidence that students' observations tend to be consistent with their predictions. There are protocols in which students' observations are clearly in conflict with prediction. In these cases, students often hedge in their observations. For example, one student predicted that the heavier object would fall faster and wrote in his observations that indeed the heavier one had fallen faster but expressed surprise that the speeds were so nearly the same.

Our observations of middle-school students, immediately after the administration of the DOE task with the two falling objects and during the eight weeks of instruction that followed, support the hypothesis that the students' interpretations of science experiments are influenced by their beliefs about the relationships between an object's mass and its speed in free fall. When the administration of the task was completed, one group of students was told that they could experiment on their own with the equipment used in the demonstrations. Several students did so. One student weighed the two blocks. He had predicted that the metal block would fall faster than the plastic block, but his observations contradicted his prediction based on the belief that heavier objects fall faster. To resolve this contradiction, he reasoned that, despite the difference in the materials from which the two blocks were made, they must be the same weight because they fell at the same rate. He proceeded to test this hypothesis by weighing the two blocks.

Two other students (Tammy and Mary) who also predicted that the heavier object would fall faster, experimented by dropping the blocks from a greater distance above the floor than the demonstrator had dropped them from. Tammy climbed onto a table and dropped the blocks while Mary put her head on the floor to hear when they struck the floor. These students reasoned that the blocks had, in fact, fallen at different rates, but that the difference in descent times was too small to be observed over the short distance (approximately one meter) used in the original demonstration. They tested this hypothesis by designing an experiment which used a more sensitive procedure (i.e., dropping the blocks from a greater height) to illuminate any difference in the blocks' rates of fall.

The observations in the DOE task with two falling objects of different mass continued to be questioned by the students throughout the eight weeks of instruction. Some of the students continued to argue that they failed to observe differences in the rates of fall because of the insensitivity of the experimental procedure. Their experiments were designed to demonstrate differences in the rate of free fall. This was to be accomplished by increasing the sensitivity of the experimental procedures by increasing the distance through which the objects dropped. They also placed their heads on the floor and either sighted along the floor or listened for the impact in order to better distinguish between times of impact. It must be remembered that these experiments, far from being idle exercises for the students, reflected a real dilemma that they faced. The students' Aristotle-like conceptions that the greater downward force applied to the heavier object implies that its speed of fall must be greater are incompatible with the observation that the objects fall at the same rate. One way to resolve this dilemma is to design experiments which will discredit that observation. A discredited observation does not act as a challenge to the students' pre-instructional conceptions, which they are reluctant to change anyway.

These observations are consistent with findings emerging from cognitive psychology that demonstrates the impact of existing knowledge in memory on the comprehension of text (Anderson, Reynolds, Schallert, & Goetz, 1977; Bransford & McCarrell, 1974; Lindsay & Norman, 1972). Early studies in the area of reading comprehension aimed at demonstrating that something other than the linguistic structure of a sentence is required to explain a person's comprehension of that sentence. The "something other" is described as the person's world knowledge and often is characterized as a "schema," "plan," or "script." Bransford and McCarrell (1974) review studies which indicate that the process of understanding text involves creation of "semantic descriptions" that use both the reader's world knowledge and the sentence input. Anderson, Reynolds, Schallert, & Goetz, (1977) indicate that an individual's "private" representation of the world can affect text comprehension. In general, studies of text comprehension indicate the facilitative effect of schemata or world knowledge. However, studies of science learning, reviewed by Driver and Easley (1978), indicate that world knowledge may be logically antagonistic to science content and may persist after science instruction.

Research we have conducted (Champagne, Klopfer, & Anderson, 1980) is consistent with the Driver and Easley results. Our work illustrates that the belief that heavier objects fall faster than lighter objects is not easily changed by instruction, thus demonstrating the strong influence that prior knowledge has on the effectiveness of instruction—in this case, the prior knowledge has an inhibiting effect on learning. In a study of beginning college physics students, about four students in five asserted that, all other things being equal, heavier objects fall faster than lighter ones. These results were particularly surprising since about 70% of the students in the sample had studied high school physics, some for two years. A chi-square test showed that students in the sample who had studied high school physics did not give significantly different responses from those who had not. In a report of a similar study of the knowledge of gravity possessed by beginning first-year physics students at Monash University, all of whom had successfully completed two years of high school physics, Gunstone and White (1981) conclude: (1) "students know a lot of physics but do not relate it to the everyday world"; and (2) "In many instances, the students used mathematical equations to explain predictions, though often inappropriately, which indicates that they had lots of physics knowledge to hand but were unskilled in seeing which bit applied to the given situation" (p. 299). In their discussion, Gunstone and White note that, not only has the students' pre-instructional general knowledge of the world been insufficiently integrated with their school physics knowledge, but the pre-existing declarative knowledge continues to dominate their explanations of physical phenomena even after instruction.

Characteristics of the declarative knowledge of beginning physics students have been compiled from the analysis of empirical studies investigating students' pre-instructional conceptions of the causes of motion of objects. These characteristics include

1. Concepts are poorly differentiated. Terms used in classical mechanics are used in everyday life—terms such as acceleration, momentum, speed, and force. The meanings of these terms as used by physicists are quite different from the way in which thay are used in everyday life. Thus, we observe that students misunderstand physics instruction because they interpret their physics lectures and texts in the context of their real-world definitions of the terms. For example, students use the term speed, velocity and acceleration interchangeably; the typical student does not perceive any difference between two propostions such as these: (a) The speed of an object is proportional to the [net] force on the object; (b) The acceleration of an object is proportional to the [net] force on the object.

2. Propositions are imprecise. This imprecision derives from at least three different sources. (a) Some of the imprecision of propostions is attributable to the meanings students have for technical concepts wich are different from the canonical meaning. Example: More force means more speed. (b) Other imprecision can be interpreted as errors of scale (Gunstone & White, 1981). Example: Gravity pulls harder on objects that are closer to the earth. (This propostion, in the context of an object falling a distance of three meters, is correct only in theory because the different in the force of gravity (approximately a part in 10^{13}) is too small to measure. If, however, the difference in distance from earth were large (several hundred kiometers), the difference in the force of gravity is significant.) (c) Other propostions are just wrong and may arise because of students' attempts to inappropriately formulate general rules of motion from their experiences in the real world. Example: Heavy objects fall faster than lighter objects.

3. Explanatory schemata are situation-specific, suggesting that there is no naive abstract representation extant in the schemata that makes the schemata applicable to large number of physical situations. In our work with

middle-school students, we have observed that they fail to notice that a proposition which they have used to explain motion in one situation is directly contradicted by a proposition they use to explain motion in another situation. This suggests that the students are unaware of any need for consistency across situations. For example, students do not recognize that the same physical laws apply to objects in free fall and to objects sliding down an inclined plane.

The research reviewed here has enabled us to develop a description of uninstructed students' declarative knowledge about the motion of objects. We now turn to research that provides descriptions of the declarative knowledge of expert physicists and novices. Having these descriptions, we can then compare the experts' and uninstructed students' knowledge states and derive specifications of the changes that instruction should promote.

THE PHYSICS KNOWLEDGE OF EXPERT PHYSICISTS AND NOVICES

Much of the information about the physics knowledge of expert physicists and novices has been obtained from the analysis of thinking-aloud protocols obtained as individuals solve physics problems. The work of Chi, Feltovich, and Glaser (1981) is an example of this approach.

In summarizing the several experiments in their study, Chi *et al.* conclude that the following differences characterize the knowledge that expert and novice physicists apply in the solution of physics problems:

1. The problem-type schemata of experts are based on physical principles (for example, energy conservation and Newton's second law) and those of novices are based on physical objects (for example, springs and inclined planes) and constructs (for example, friction and gravity).
2. The contents of the schemata of experts and novices do not differ significantly in information content; however, the novices' structures lack important relations, specifically, relations between the surface features of the problem and the scientific principles which are the basis for solutions.
3. Links exist between the experts' abstract representation of features of the problems and the physical principles which are the basis for the solution of the problem. These links do not exist in novices.
4. Experts' schemata are arranged hierarchically along the dimension of abstractness; in contrast, the different levels of the novices' knowledge are not well integrated, thus preventing easy access from one level of abstraction to another.

Summary

Significant differences in knowledge exist about motion possessed by the uninstructed person, the novice, and the expert physicist (Champagne, Gunstone, & Klopfer, 1982). The differences associated with the quantity, accuracy, abstractness, and structure of the information has important implications for learning and instruction. The differences help explain observed difficulties students have in learning mechanics and provide useful information for the design of instruction.

Empirically derived descriptions of the characteristics of the knowledge of the motion of objects of uninstructed students, novices, and experts are summarized in Table 5.1. Although this summary table cannot display all the nuances associated with the knowledge, it does make evident the contrasts and similarities in the characteristics of the three groups' knowledge with respect to principles, surface features, and second-order features, each of which is briefly explained in the first column. The second column in the table describes current conceptions of the characteristics of physics knowledge of the uninstructed students. The fourth column describes characteristics of the experts' declarative knowledge. By comparing the characteristics of the uninstructed students' and experts' knowledge, the nature of the needed changes in the students' knowledge can be specified.

It should be emphasized that, when the instructional strategies we shall propose are employed, we do *not* expect the students' knowledge to pass through a transition stage showing all the characteristics of the novices' schemata described in the third column of Table 5.1. Specifically, our proposed instructional strategies will encourage the reconciliation and integration of the uninstructed students' existing knowledge developed from real-world experiences with the knowledge developed as a result of formal physics instruction. Thus, the creation of the conflicting knowledge observed in many novices is avoided.

COGNITIVE OBJECTIVES AND STRATEGIES FOR MECHANICS INSTRUCTION

One way of conceiving of the objectives of instruction is in terms of changes which will transform students' existing knowledge into an approximation of the knowledge of experts. By contrasting specific features in the knowledge states of experts and uninstructed students (and sometimes of novices), we can derive cognitive objectives to be attained by the student with respect to those features.

Table 5.2 summarizes essential differences in the declarative knowledge of the three groups and the instructional objectives for mechanics instruction

TABLE 5.1 Domain-Specific Knowledge of Uninstructed Students, Novices, and Expert Physicists

Features	Characteristics of Unstructured Student's Knowledge	Characteristics of Novice's Knowledge	Characteristics of Expert's Knowledge
Principles Ideas of some degree of generality that express relationships; principles are applied to solving problems; can serve to organize schemata.	Principles are generalized rules derived from everyday experiences (world knowledge). They are imprecise propositions. The imprecision is due to vagueness about the meaning of concepts, errors of scale, and inappropriate formulations of general rules. The principles (rules) have limited scale and tend to be situation-specific. The notion that an abstract principle can apply to a range of different physical situations is lacking or poorly developed. There appears to be no awareness of the need for consistency among the rules that cover different physical situations.	Principles are relationships between physical variables expressed as equations or rules. Some of the principles are the major physical laws expressed in equation form, but there is no evidence that they serve as organizers of schemata.	Principles are major physical laws, which are highly abstract and express relationships of great generality. Included with each principle are the conditions under which the principle applies. Principles have associated schemata, which are oriented by the content and applicability conditions of the principle. The applicability conditions usually are expressed in terms of second-order features.
Objects and Physical Conditions (Surface Features) Physical objects and conditions described or presented in a problem situation; physical features of objects and their states of motion or position that are directly perceivable from verbal description; diagrams or direct observation of the physical situation.	Concrete objects and the directly observable properties of objects are present. A reasonable inference is that the objects and properties define the specific physical situation which, in turn, directs the search in memory for a general rule that covers it.	Physical objects and their surface features are the basis for categorizing problems. It is inferred that an object or a configuration of objects functions as the organizing element (node) in its schema for the problem. The content of the representations may be concrete objects or abstractions at the level of diagrams.	Concrete objects, their physical configurations, and diagrams of objects are present in the schemata, but none of these is prominent. It is inferred that objects serve primarily as vehicles for identifying second-order features, and that sometimes they trigger the activation of a particular principle-based schema.
Physics Concepts and Symbols (Second-Order Features) Idealizations of physical objects (e.g., an elephant is represented as a point mass) and constructs or entities (e.g., energy, force); conventional representations of physical entities (e.g., vector components).	There is no evidence that second-order features are represented in these schemata. Concepts and terms are present, but many are poorly differentiated. The meanings of the terms are their real-world meanings, rather than their technical meanings in physics.	Some conventional representations of physical entities are present, and idealizations of physical objects may be used in problem representations. Concepts and terms related to the objects which dominate the problem-solution schema are present. Novices report taking terms directly from the problem statement to identify equations that could be approximately employed in solving the problem.	Representations of physical objects in their idealized form are prominent, with the content of the representations determined by the organizing schema. Physical entities are represented according to the conventions of the field. Some experts report that features abstracted from problem statements help to select the basic approach to problem solutions, thus indicating direct links between these second-order features and principles. Concepts relevant to the organizing schema are present. Associated with each concept are its interconnections of relationships with other concepts and with the schema's major physical law.

TABLE 5.2 Contrasting Features in Declarative Knowledge and Their Related Instructional Objectives and Strategies

Contrasting Features in Declarative Knowledge States of Unstructured Students (U), Novices (N), and Experts (E)	Instructional Objectives Derived from the Contrasts	Instructional Strategies to Facilitate Attainment of Instructional Objectives
CONCEPT MEANING Meanings attributed to technical terms by U differ in significant ways from the meanings of E and N. Example: U - acceleration means speeding up; E and N - acceleration means a change in the magnitude or direction of velocity.	Students know both the everyday meaning and the canonical definition of mechanics and can specify differences between the everyday and canonical meanings.	*Interactive dialogue using multiple instances:* Provides students with opportunities to become aware of the meanings they attribute to physical concepts, how these meanings differ from concept to concept. Examples: (1) doing physics problems or describing a physical event to a friend, or (2) doing mechanics problems in which there is no motion. Dialogue provides students with opportunities to contrast their meanings of concepts with those of the physicists.
CONCEPT DIFFERENTIATION U do not differentiate mechanics concepts. Example: U - *weight* and *mass* are the same thing; N and E - *mass* and *weight* are perfectly correlated but distinct.	In analyzing a given physical situation, students can explain which of two poorly differentiated concepts is the relevant concept to apply.	*Interactive dialogue using multiple representations:* Provides students with opportunities to verbalize their analysis of physical situations in a way that simply substituting 9 g for mass (m) or 45 dynes for weight (F) in an equation does not. We hypothesize that the verbalization will help differentiate weight (a force) expressed in dynes from mass expressed in grams.
PROPOSITIONS IN SCHEMATA U - presence in schema of incorrect propositions. Example: motion implies force. N - presence in schema of conflicting propositions which are applied in different situations. Example: Motion implies force proposition applied in the analysis of real-world situations. Change in motion implies force proposition applied in the quantitative solution of physics problems. E - propositions present in schema are internally consistent and widely applicable. Example: Change of motion implies force proposition is applied in analyzing all pertinent problems.	Students apply change in motion implies force proposition in real-world situations. Students contrast implications of the difference in the two relationships between force and motion expressed by the propositions (1) Motion implies force, and (2) Change in motion implies force.	*Ideational confrontation:* Provides opportunity for students to (1) be explicit about the propositions they assume in invoking the presence of forces in physical situation (For example, U generally invokes forces only in situations where there is motion.) and (2) make explicit the relationship between motion and force in the propositions they use.
STRUCTURAL FEATURES Concept integration of U and N is sparse, with fewer links among concepts than for E. Example: N - experientially derived motion-of-objects schema is not integrated, or reconciled with Newtonian mechanics schema. Integration of representations in U and N is poor, while they are well-integrated in E. Example: N - representations of surface features of physical situations are poorly integrated with abstract representations of physical situations; these, in turn, are poorly integrated with propositions which link canonical objects and physical constructs used in abstract representations; E - representations of surface features of physical situations are integrated both with abstract representations of physical situations and with propositions linking the canonical objects and constructs of abstract representations.	Students qualitatively analyze mechanics problems: (1) Produce an abstract representation of a physical situation. (2) Recognize that situations with very different surface features can have the same abstract representation. (For an example, see Appendix A where the situations of 6 problems have the same abstract representation.) (3) Recognize that the problems can be solved by the application of the same mechanics principle.	*Qualitative analysis of problems to change structural features of mechanics schema:* Forges links between the physical situation, its abstract representation using canonical objects and mechanics constructs, and the principles (Newton's second law, F = ma) which link properties of the canonical objects and constructs. Also forges links between concepts (e.g., between *mass* and *weight*) to integrate them better, thereby contributing to concept differentiation.

based on those differences. These objectives specify features of the ideal state which the learner will approximate but do not detail how far the learner will move along the continuum from beginning student to expert as a result of a particular course or sequence of instruction. Further refinement of the objectives for a certain instructional sequence must take into account many other factors, including the content the instruction will cover, the time available for instruction, and the age and academic aptitude of the students for whom the instruction is intended. Also included in Table 5.2 are instructional strategies proposed to achieve the objectives.

The process of specifying instructional strategies based on differences in declarative knowledge is more complex than that of objectives specification. The process requires the analysis of traditional formal physics instruction, informal educational experiences, and everyday experiences with moving objects—in short, analysis of the many situations that affect the development of knowledge about the motion of objects. Based on available information concerning differences in the experiences of the three groups (uninstructed students, novices, and experts), it is possible to hypothesize how differences in experiences produce the observed differences in declarative knowledge. This process is informed by theory and the results of empirical studies investigating the relationship between experience (usually formal instruction) and cognitive change.

The following example relates differences in declarative knowledge with differences in experiences. The three groups of interest differ with respect to (1) the quantity and extent of formal mechanics instruction, (2) experience in solving mechanics problems, and (3) the extent of their verbal interactions about mechanics. The studies by Chi et al. (1981) suggest that the contents of novices' declarative knowledge pertinent to particular types of physics problems are similar to those of experts with respect to objects, concepts, and terms; however, experts' knowledge structures contain many more linkages. (It is important to this discussion to note that the more extensive information base, which experts have acquired as a result of their greater exposure to formal instruction, is not necessary for and indeed does not seem to be called upon in the successful solution of problems of the type on which the analysis of expert-novice differences is based.) The observation that novices' cognitive contents resemble those of experts is an indication that the didactic method of instruction is effective for imparting discrete bits of information. However, the detailed analysis of problem-solving behaviors of novices suggests that the structural organization of the information resulting from exposure to didactic instruction is less than satisfactory. This suggests an important limit on the effectiveness of didactic instruction. We hypothesize that the multiple *links* between the contents of experts' knowledge structures and their more abstract schemata develop, in part, from extensive practice in problem solving. Cognitive theory suggests that the development of highly integrated, abstract declarative knowledge structures may be facilitated by verbal interactions with others about mechanics.

A significant difference between the physics experience of experts and novices is that experts have had greater opportunities for communicating about physics with other people. This suggests that verbal interaction may have an effect on building the desired links between the contents of students' knowledge structures that parallels the effect of extensive practice in problem solving. In theoretical support of this notion, Anderson (1977) has suggested that one type of verbal interaction, namely, Socratic dialogue, is a possible mechanism for producing schema change. We proposed several instructional strategies that involve dialogues to bring about the students' attainment of the kinds of objectives identified in Table 5.2. Although student-student and student-teacher dialogues are the common denominator in these strategies, we distinguish several dialogue strategy varieties on the basis of differences in instructional procedures, in the stimulus materials used, and in the instructional objectives which the strategies are designed to further. Four of these dialogue-based strategies represented by the strategies in the rightmost column of Table 5.2 are interactive dialogues using multiple instances, interactive dialogues using multiple representations, ideational confrontation, and qualitative analysis of problems. We shall describe and illustrate each of these instructional strategies.

INTERACTIVE DIALOGUES USING MULTIPLE INSTANCES

This strategy is conceived to facilitate development of (1) physics problems schemata and (2) correct physical meanings of physics terms. An illustration of the procedures of the multiple instances strategy follows. The students first are presented with a set of five or more mechanics problems which require qualitative answers. Two sample problems taken from a typical set are shown in Figure 5.1. These problems are qualitative restatements of problems from different physics textbooks. The physical situations or surface features in both the qualitative and quantitative versions of all the problems in the set are very different, but all the problems can be represented in the same abstract form (diagram, algebraic symbols, or verbal description using mechanics concepts) and can be solved using the same mechanics principle. Each student produces a solution to each of the problems in the set and then shares with the class the problem analysis, the solution of the problem, and the definitions of technical terms used in the solution or the analysis. This procedure forces students to be explicit about the idiosyncratic meanings

QUANTITATIVE VERSION

3. Rifle and Bullet Problem

A 3-g bullet is fired from a 2.4-kg rifle with a velocity of 360 m/s north. Find the momentum of the bullet and the recoil velocity of the rifle, assuming that no other bodies are involved.

(Smith & Cooper, 1979, p. 93)

5. Carts and Spring Problem

Two heavy frictionless carts are at rest. They are held together by a loop of string. A light spring is compressed between them (see drawing). When the string is burned, the spring expands from 2.0 cm to 3.0 cm, and the carts move apart. Both hit the bumpers fixed to the table at the same instant, but cart A moved 0.45 meter while cart B moved 0.87 meter. What is the ratio of:

(a) The speed of A to that of B after the interaction?

(b) their masses?

(Haber-Schaim et al., 1976, p. 321)

QUALITATIVE VERSION

A bullet is fired from a rifle. Describe the motion of the rifle. How does the velocity of the rifle compare with the velocity of the bullet?

Two heavy frictionless carts are at rest. They are held together by a loop of string. A light spring is compressed between them. The string is burned and the spring expands. Describe the motion of the carts. How does the velocity of cart A compare with the velocity of cart B?

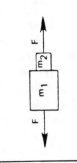

PHYSICIST'S ANALYSIS

Forces of equal magnitude and opposite in direction are exerted on two unequal masses at rest. How do the velocities of the masses compare? How do the displacements of the masses compare?

FIGURE 5.1 Physics problems with different surface features and the same deep structure.

attributed to technical terms and the principles and propositions that they apply in the analysis of the problem. Each student can contrast his or her analysis of a problem with the analyses presented by other students.

When all of the problems in the set have been considered by the class, the teacher presents the physicists' analysis of the problem by means of diagrams and verbal explanations using the technical vocabulary of mechanics. The expert analysis, based on the common deep structure of all of the problems in the set, is shown in Figure 5.1. The teacher demonstrates that the abstract representation is the same for each of the problems and that the same principle produces a solution to all the problems in the set. Then students analyze their solutions so the problems in light of the physicists' solution and specify how their interpretations differ from that of the physicists.

Our illustrative outline of the procedures used in the multiple instances stragegy concerns interactive dialogues about just one set of problems. To be effective in producing the desired change, of course, the strategy must be repeated with other sets of problems with the same deep structure.

INTERACTIVE DIALOGUES USING MULTIPLE
REPRESENTATIONS

The optimal application of this instructional strategy is in situations where the students need to develop understanding of a concept as it applies in a given class of situations. This instructional strategy requires that students work with various representations of the concept, including diagrams, three dimensional manipulative objects, physical models, conventionalized symbols, data tables, and graphs in a way that makes explicit the correspondences among the various representations. We illustrate the strategy by outlining a portion of an instructional sequence. The example involves a 1.0 kg block placed on an inclined plane at various angles ($0°-90°$) and being held in dynamic equilibrium by a force parallel to the incline (see Figure 5.2). In this situation, three forces are acting on the block—the gravitational force (F_g), the force parallel to the incline (F_i), and the normal reaction force (F_n). One aspect of the importance of this situation in elementary dynamics is that an understanding of the conditions necessary for dynamic equilibrium enables knowledge of the F_n and F_i to be used to establish the magnitudes of the components of F_g parallel and perpendicular to the incline. The multiple representation strategy in which a single physical situation is represented at different levels of abstraction helps force important mental links between the physically relevant information in a physical situation and the latent information in the scientist's abstract representations of the situation.

As shown in Figure 5.2, the physical apparatus to be used in this example consists of a board (the inclined plane), a block (the 1.0 kg object), and two

PHYSICAL
SITUATION

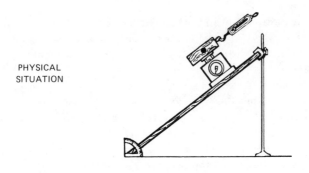

SYMBOLIC REPRESENTATIONS

1. Data Table
(numerical and verbal representations)

Angle of incline	45°
Platform scale reading	6.9N
Spring scale reading	6.9N
Weight of block	9.8N

5. Algebraic-Trigonometric Equations
(algebraic and verbal representations)

Angle of incline	θ
Normal reaction force	$F_n = F_g \cos \theta$
Force parallel to incline	$F_i = F_g \sin \theta$
Mass of block	m
Gravitational acceleration	g
Force of gravity on block	$F_g = mg$

2. Vector   3. Parallelogram Vector   4. Right Triangle Vector

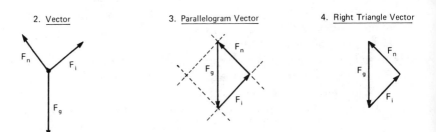

FIGURE 5.2   Multiple representations of an equilibrium condition.

spring scales—one a compression spring platform scale to measure the normal reaction force, the other an expansion spring scale to measure the force parallel to the incline. There are six symbolic representations to be coordinated with the physical situation. One is the numerical representation in a data table, and three more are vector representations. One vector representation shows three forces acting through the block's center of mass. Two others show the vectors rearranged for analysis, one by construction (the parallelogram representation), and the other by the trigonometric relationships of the right triangle. The fifth representation is algebraic-trigonometric and consists of equations for the principal forces in the situation. While all of these representations are equivalent mathematically, conversion from one to the other can be puzzling to the students, who may not recognize the mathematical equivalence. Finally, examples of the sixth symbolic representation are shown with both the numerical representation in the data table and the algebraic-trigonometric representation. Here verbal representation is used to identify measured quantities and the forces acting on the block.

It is by no means a straightforward task for students to coordinate the various symbolic representations with each other and with the physical situation. The instructional strategy aims at mapping the information available in the physical situation onto the symbolic representations and at making the correspondences among the several representations explicit. This is accomplished by having students work simultaneously with two representations and noting how changes in one (for example, an increase in angle of incline and concurrent changes in the two spring scale readings) result in corresponding changes in a symbolic representation (for example, an increase in the angle between F_g and F_n; a lengthening of the vector representing F_i and shortening of the vector representing F_n). Correspondences among the symbolic representations are similarly established by manipulating a variable in one (a change in magnitude of the angle of incline) and mapping the corresponding changes on another symbolic representation or on the physical apparatus.

IDEATIONAL CONFRONTATION

The instructional dialogue strategy which we call "ideational confrontation" is designed to help bring about substantial changes in the content of students' motion-of-objects schemata. Certain concepts and propositions in the uninstructed students' schemata must be replaced by or integrated with the concepts and propositions of the physics experts' schemata. We can illustrate this by describing certain specific changes we anticipate in students' motion-of-objects schemata.

Figure 5.3 displays three representative problems which students in an introductory physics course would confront. The first problem is to make a prediction comparing the amount of time required for two blocks of unequal mass to fall a distance of 1 meter. Our hypothesized falling-objects schema

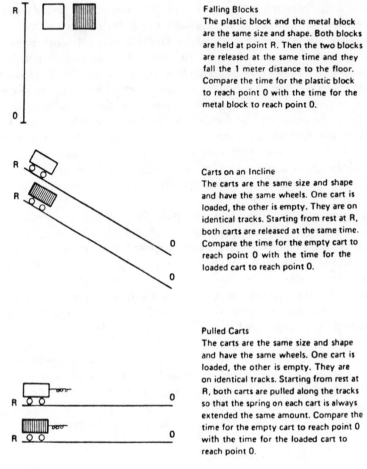

Falling Blocks
The plastic block and the metal block are the same size and shape. Both blocks are held at point R. Then the two blocks are released at the same time and they fall the 1 meter distance to the floor. Compare the time for the plastic block to reach point 0 with the time for the metal block to reach point 0.

Carts on an Incline
The carts are the same size and shape and have the same wheels. One cart is loaded, the other is empty. They are on identical tracks. Starting from rest at R, both carts are released at the same time. Compare the time for the empty cart to reach point 0 with the time for the loaded cart to reach point 0.

Pulled Carts
The carts are the same size and shape and have the same wheels. One cart is loaded, the other is empty. They are on identical tracks. Starting from rest at R, both carts are pulled along the tracks so that the spring on each cart is always extended the same amount. Compare the time for the empty cart to reach point 0 with the time for the loaded cart to reach point 0.

FIGURE 5.3   Three representative introductory physics problems.

for a typical uninstructed student includes the following features. Considering first the object category, the schema contains objects together with their observable physical properties, such as size, shape, and weight. Among the physics concepts in the schema are the object's properties, including volume, mass, and weight, though these concepts are not precisely defined and are poorly differentiated; the objects' average speed in falling and the instantaneous speed at any point (two poorly differentiated concepts); the time elapsed during the fall; and a downward pull on the objects due to the pull, or force, of gravity. Here also is the concept, heavier than, in the sense of greater weight, and the specific application of this concept to the equal-sized objects of different materials with the result that the metal block is conceived to be heavier than the plastic block. Relevant to this conception is the general rule, stored among the schema's experience-derived propositions, that "Heavier objects fall faster than lighter objects." The uninstructed student applies what he or she believes to be the relevant general rule to the specific physical situation to deduce an answer for the problem. For the situation of the two blocks, the reasoning is: (1) The metal block is heavier than the plastic block. (2) Heavier objects fall faster than lighter objects. (3) Hence, the metal block falls faster than the plastic block. (4) The metal block takes less time to reach point 0 than the plastic block. The form of reasoning leading to the problem's solution is the same, of course, if the relevant general rule selected from the schema's stored propositions is: "According to Galileo, all objects fall at the same speed regardless of their weight." Howeverer, the answer to the problem now is that the metal and plastic blocks reach point 0 at the same time.

To transform the hypothesized schema of the uninstructed student into a schema approximating an expert's schema requires changes such as the following. With regard to concepts in the schema, in place of concepts which are vaguely defined and poorly differentiated, the transformed schema's concepts have precise qualitative definitions which distinguish them one from the other. For instance, the concepts of *volume, mass,* and *weight* are differentiated from one another, and the definitions of *mass* and *weight* incorporate the relationships of these concepts to each other and to major physical principles. In a similar position in the transformed schema is the concept of *force,* a generalized concept central in several major physical principles that was formerly represented only as a *push* or *pull* concept. With regard to principles in the student's new schema, in place of imprecise, experience-derived rules of limited scope and applicability, the transformed schema is dominated by a major physical law which is abstract and broadly applicable, specifically, Newton's second law of motion. Associated with this law are its conditions of applicability, including that the idealization assumptions of Newtonian mechanics are realized, and that a constant net unbalanced force is acting on the objects of interest, here the plastic and metal blocks. Recognition that Newton's second law is applicable to this situation requires that the student's new schema include appropriate links between the abstractions of Newton's second law (mass, force, and acceleration); the physical objects (plastic and metal blocks, identical in shape and volume but differing in mass) and states specified by the problem (initial and final position, initial velocity is zero); and physical constructs not specifically mentioned or observed (force of gravity, acceleration, air resistance). These abstractions or second-order features have no parallel in the student's initial schema and must be abstracted from the physical situation and the real

Original Problem

"9.   A freight engine of mass 20,000 kg accelerates from rest up to a velocity of 2.0 m/sec. in 5.0 sec. If it is pulling a train of 20 cars, each with a mass of 10,000 kg, what is the force in the coupling between the engine and the first car?"

(Stollberg & Hill, 1964, p. 49)

Problem Restated in Qualitative Form

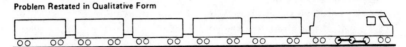

A freight train is stationary. It then accelerates away from rest. While the train is accelerating, how does the force exerted by the engine to accelerate the train compare with the force in the coupling between the engine and the first car?

Subsidiary questions: (i)   What assumptions did you make to arrive at your answer?
(ii) What principles of physics/laws of motion did you apply to the situation in coming to your answer?

Knowledge/skills assumed in following outline of Instructional Dialogue strategies:

    1. It is assumed that students
       (a) have completed a study of kinematics and Newton's Laws
       (b) are aware that the extension of a rubber band or spring is a reasonable indicator of the magnitude of the extending forces acting on the rubber band or spring.
    2. It is also assumed that there is instructional value in students having both visual and touch inputs of estimations of the magnitude of forces.

Strategy A   Outline of strategy to be used for responses to the qualitative problem of the form "don't know" or "the same."

Steps in the Strategy	Purpose of Steps	Commentary on Steps [1]
A1. Have the student pull, using a rubber band, laboratory trolleys (PSSC) carrying various numbers of bricks.	To draw attention to, or re-establish, the qualitative relationship between mass, motion of the mass, and applied force.	Direct observation. Questions to assist processing of the observations would follow, for example, Collins (1977) Rule 1: Ask about a known case, and Rule 11: Specify how the variable depends on a given factor. Specific procedural rules will vary between individuals.
A2. Use two trolleys connected by a rubber band arranged so that the trolleys are almost touching when stationary — hence separation of trolleys is an indicator of extension of that rubber band.	(a) To establish what specific objects each of the two rubberbands is pulling. (b) To direct attention to the qualitative relationship between mass and force required to accelerate that mass.	As for A1 above. Possible procedural rules include Collins (1977) Rules 13: Probe for differences between two cases; and 11 (see above).

Have the student pull the trolleys by a rubber band attached to cart 1.

FIGURE 5.4   The freight train problem.

Steps in the Strategy	Purpose of Steps	Commentary on Steps
A3. Repeat A2 with various masses added to cart 1 and/or cart 2.	See purpose (b) in A2.	See A2 above.
A4. Repeat A2 and A3 using a number of carts.	See purpose (a) & (b) in A2.	See A2 above.
A5. Return to qualitative problem. After successful solution of the problem as asked, use strategy B below, if appropriate.		

Strategy B   Outline of strategy to be used for responses to the qualative problem of the form "force exerted by the engine is greater than the force in the coupling" (i.e., correct response to question asked).

Steps in Strategy	Purpose of Steps	Commentary on Steps
B1. Ask student if they can be more precise about the relative magnitudes of the two forces.	To establish that ratio of forces is inverse of ratio of total mass/weight of train to mass/weight of cars pulled by the engine. If this is not forthcoming, go to strategy A, beginning at A2. If some such statement is produced, go to B2.	Collins (1977) Rule 4: Ask for prior factors.
B2. Repeat subsidiary questions (i) and (ii) from the qualititive problem.	To link answers to the subsidiary questions to the specific stiuations presented — both the experiences with carts and the qualitative problem.	Collins (1977) Rule 2: Ask for any factors. Also, in the exploration of individual responses, one or several applications using other rules may be made.

[1]Note. Collins (1977) proposed a series of procedural rules for this form of instructional dialog, and we comment on particular rules which are pertinent to the several steps in our strategy. The procedural rules were recognized and amplified in Collins & Stevens (1981).

FIGURE 5.4   *(continued)*

objects therein, in order for the student to link the physical situation with the second law. Examples of other second-order features are canonical objects, all of whose mass is concentrated at a single point, and vector representations of forces. Given that these second order features are linked with physical objects, the three "real-world" situations described in the problems in Figure 5.3 can all be represented abstractly as pairs of unequal masses (two blocks, two carts), each under the influence of a constant net force (gravity, extended rubber bands). Assuming the influence of friction is minimal, Newton's second law allows the student to predict constant acceleration for each of the objects and to compare the magnitude of the acceleration produced by unequal forces on unequal masses in the falling blocks and inclined plane problems and on unequal masses under the influence of equal forces in the pulled carts problem.

Despite the numerous changes in the student's schema that we have surmised, some characteristics of the initial schema need not and should not change. In this category are the schema's representations of the physical objects themselves, their properties, and their described or suggested behavior. These are the representations that are utilized, we believe, to initiate a qualitative analysis of a problem. Hence, they are important in the application of our fourth instructional strategy.

QUALITATIVE ANALYSIS OF PROBLEMS

When an instructional objective calls for altering the structural organization or integrating the concepts and propositions of a schema, the instructional dialogue strategy of the qualitative analysis of problems can be very useful. In order to explain and illustrate the procedures for the qualitative analysis of problems, we have analyzed a number of prototypical physics problems taken from introductory texts. An example of such an analysis is shown in Figure 5.4. Other examples can be found in Champagne, Klopfer, and Gunstone (1982). The general structure of each analysis is simple. Initially, the standard form of the problem is rewritten in qualitative form, with subsidiary questions asking about limitations of, and physics principles used, in the answer given by the student. Then a realistic minimum state, in terms of relevant prior knowledge and experience is selected, and a strategy for working from the state to a successful solution is proposed. At this stage, when we do not have the insights to be derived from data, it is assumed that other, more developed responses can be accommodated by beginning at a later point in this sample strategy. The strategy has been outlined only. It indicates a series of logical steps. Within each step, the essential concept(s) to be developed and the purpose of the step in terms of the problem solution are indicated. In some cases, a particular instructional methodology to be used for a step is shown, while in the remaining cases, only instructional dialogue is anticipated. For each step, the rationale for the procedure is described, referring to the techniques identified by Collins (1977) and Collins and Stevens (1981) when appropriate. For each of the three examples, a procedure for handling a correct answer to the questions asked so as to develop an approach to solving the problem is also given.

PHYSICS PROBLEMS AND NUMERICAL EXERCISES

The development of our ideas concerning the qualitative analysis of physics problems has led us to a simple but, we believe, most important reconceptualization of the nature of physics problems vis-a-vis the structure of standard physics texts and curricula. In what seems an obvious extension of the form of solution strategy experts apply to standard physics problems, we wish to describe qualitative problems of the form given here as "Physics Problems." The solution strategy derived from such a problem can then be applied to a set of exercises of the form currently termed problems. These exercises we wish to describe as "numerical examples." The parallels with mathematics instruction are striking. The normal mode of instruction in math involves the development (or, unfortunately, the provision) of a solution strategy for a particular type of problem, for example, factorization of quadratics. Then students are asked to apply this solution strategy to numerical examples. These numerical examples are labeled by most texts and teachers with a description such as "examples," and "exercises," but are rarely termed "problems."

This reconceptualization has, we believe, considerable instructional significance. Current practices seem to be based on the assumption that, by doing "numerical examples," students will themselves extrapolate solution strategies to "problems." Texts commonly approach the task of elucidating methods for solving standard physics questions by giving "numerical examples" and, starting immediately with an equation, describing the appropriate approach. It is apparently assumed that this approach will be transferred by students to other appropriate "numerical examples," but instructional assistance for this transfer is rarely given. Indeed, as any practitioner of physics teaching can testify, assisting students to make such transfer is extraordinarily difficult. As a consequence, we argue that the methods outlined here are designed to foster the development of problem solutions. We believe that solution of numerical examples is a subsequent and more simple step.

INSTRUCTIONAL ISSUES

Two issues which merit some consideration here arise from the instructional strategies we are advocating. First, our proposed instructional dialogue strategy focused on the qualitative analysis of physics problems that makes use of laboratory equipment in some cases (e.g., Step A2, Strategy A in Figure 5.4). This use of laboratory-type experiences may be questioned since laboratory work has not been shown to produce significant cognitive gains for science students. Hence, it is necessary to discuss the specific conceptual basis for our use of laboratory-type experiences in the instruction. The second instructional issue to be discussed is related to our proposed multiple representations strategy and concerns the proper role of analogies in science instruction.

Role of Laboratory Experiences

Studies of instruction with laboratory work versus instruction without laboratory work have consistently suggested no cognitive advantage for the lab work group exists (see, for example, Bates, 1978, for a review). However, much laboratory work in the sciences, in general and physics, in particular can be characterized as follows: students follow some procedure for the collection of data, which is the primary focus of their activity while involved with the apparatus; results are then analyzed and, in many cases, subsidiary questions answered; the end result is the verification (or "discovery") of some law, relationship, or effect of one variable on another. This mode of laboratory work has most frequently resulted in students concentrating on the procedures for data collection and, overwhelmingly, on obtaining the correct answer. In such circumstances, the apparent failure of laboratory activity to markedly foster conceptual development does not seem surprising.

Gagné and White (1978, p. 214) have argued that laboratory work which is performed with the intent of facilitating understanding and conceptual development will better achieve such purposes if it seeks to generate links between episodes and images (derived from the laboratory work) and propositions and intellectual skills. This is an appealing argument, especially given the extent to which introspection reveals images and episodes in one's own memory which relate to specific facts and principles. However, Atkinson (1980), in comparing chemistry instruction with laboratory work and chemistry instruction with only demonstrations, found that the laboratory work did not appear to assist that group in recalling related subject matter even though this group was somewhat more likely to recall relevant images and episodes. Gunstone (1980), in an investigation of physics learning, found virtually no propositional connections between physics concepts which were derived from laboratory work. Since both of these studies utilized laboratory work in the verification mode described above, it is plausible to suggest that such laboratory work, because of the focus it gives to student activity, is unlikely to generate the links among images, episodes, propositions, and concepts argued by Gagné and White.

In considering the lack of cognitive gain from traditional laboratory work, White (1979) has argued for the inclusion of three types of practical activities not usually found in science courses. One of these activities is used in the strategies for developing problem solutions which are outlined in Figure 5.4. This activity

> is intended to establish generalized episodes involving materials and events of common experience, with the purposes of linking school subject matter and daily life and of providing experiences which will be called into play in making subsequent information comprehensible. (White, 1979, p. 387)

Directed hands-on experience with physical objects is included in our strategies because of the extent to which such experience can relate aspects of the problem to relevant prior experience in the world at large. It can also provide sensory inputs about situations and provide episodes or images as memory structures for later recall. It is of interest to note that this overt attempt to bring to the solution of the problem relevant real-world knowledge is in line with arguments for the importance of such knowledge to the individual's problem solution (see, e.g., Duran, 1980; Neisser, 1976). Others (e.g., West, 1980) have probed students' links between curriculum knowledge and general knowledge as an important dimension of students' cognitive structures.

This rationale for a particular form of laboratory work will be briefly illustrated by reference to the proposed strategy for the freight train problem in Figure 5.5. The use of rubber bands to pull and connect laboratory trolleys illustrates the basic principles underlying the freight train problem. By manipulating masses and the number of trolleys, students will derive *qualitative* relationships between the relevant variables, be assisted to relate these relationships to the underlying physics principles, and, perhaps most importantly, be given a memorial referent to which they may relate this information. In addition, some students will relate this experimental manipulation to previous relevant experience (e.g., pulling peers while skating). Before the event, it is clearly not possible to determine which students have relevant prior experience. Hence, the drawing out and relating of any such experience will be achieved by the instructional strategy to be used.

Seen in this light, the use of the trolleys can be clearly characterized as an attempt to provide a physical analogy to the freight train. Thus, in this instance, an analogy is used as a part of our qualitative analysis of problems strategy. Analogies crop up even more frequently in our proposed instructional strategy of interactive dialogues using multiple representations. For this reason, we need to be aware of the issue of the potential dangers of the use of analogies in instruction.

Analogies in Instruction

The role of analogies and models (which are little more than formalized or institutionalized analogies) in the development of science has been widely discussed by philosophers of science. Further, much has been written about the manner in which analogies can inhibit or distort the development of science when inappropriate deductions are made on the basis of existing analogies (Harré, 1965; Nagel, 1961; Scheffler, 1969). A parallel situation exists in science instruction as Gee (1978) and Weller (1970) have indicated, where the use of analogies can result in students making inappropriate extensions of an analogy to generate false information. White and Gunstone

(1980), in an investigation of cognitive structures associated with electricity, found a clear and extreme example of an inappropriate extension. A college science graduate who had been given the water flow analogy during his early instructional experience with electric current had extended the analogy on his own initiative so as to equate the functions of a pipe in the water analogy and the insulation surrounding a conducting wire. As a consequence, he believed resistance to electric current arose from the friction between electricity and the surrounding insulation and that electrical leakage was akin to a hole in a water pipe and that it could result from a bare electrical conductor. He had deduced that much of the research effort in electrical engineering was devoted to the production of more efficient (i.e., less friction-generating) insulating covers for wires.

On the positive side, Gentner (1980) has sought to frame the issue of how multiple representations may help to develop understanding of science concepts in terms of structural differences between good and poor analogies. Gentner argues that the explanatory power of analogies in learning science concepts rests on the fact that analogies point up equivalent relationships, rather than equivalent objects, within two distinct systems. Gentner develops a set of criteria by which to evaluate the likely effectiveness of particular analogies in explaining science concepts: clarity, a measure of the precision of the mapping; richness, a measure of the sheer number of mappings between two systems; abstractness, a measure of the level of relationships mapped between the systems; and systematicity, a measure of the extent to which the constraints in one system of the analogy are isomorphic to the constraints in the second system.

An important and neglected aspect of the instructional use of analogies is an understanding of the limitations of the analogy. Developing such understanding also adds to the understanding of the science concept which the analogy is used to illustrate. Consequently, we are explicit about significant limitations to be developed with students when using analogies in providing experiences with the multiple representations of a concept.

SOME ASPECTS OF PROPOSED RESEARCH

The foregoing sections of this chapter have explicated an alternative approach to conceptualizing the well-known instructional problem of the difficulty that beginning physics students experience in learning mechanics. We have sought to identify and characterize the entering students' alternative conceptions of motion that contravene the physicists' declarative knowledge about motion, the knowledge whose contents and structure are desired outcomes of mechanics instruction. Finally, taking seriously the students' alternative conceptions, we have argued the necessity of deliberately altering the students' declarative knowledge to align it with that of the physicists', and we have suggested several dialogue-based instructional strategies for effecting these changes. However, we are not content with proposing a new approach, no matter how logical or how well supported by research it might be. We intend to investigate the strategies and issues which have been argued above. The major hypothesis to be tested in our proposed research is that engaging beginning physics students in instructional dialogues that utilize one or more of our proposed instructional strategies will produce changes in the students' declarative knowledge about the motion-of-objects. Our further hypothesis is that, after completion of the specified instruction, the students' schemata will have several significant features which are characteristic of the declarative knowledge of physics experts.

Our first task in this research is to specify the scope and sequence of questions and manipulative experiences which will serve as the stimuli for the instructional dialogues. We have illustrated the rationales for the construction of stimulus materials in conjunction with the example exhibited in Figure 5.4. From analyses similar to these, for every strategy we select for the instruction, we can derive specifications for the issues which need to be raised, the topics which will be discussed in the dialogues, and the laboratory-type experiences which are required to provide essential observations or episodes. With the instructional strategies and the content of the instruction specified, the final design task is to incorporate these into instructional materials.

For the multiple instances strategy, for example, we plan to select problems that form a sequence in which the problems, though differing in surface features, can all be solved by making essentially the same qualitative analysis and by utilizing the same physical principle. The three physics problems shown in Figure 5.3 exemplify this kind of sequence. There will be several such sequences in the instruction, whose scope is an introduction to dynamics under the purview of Newton's laws of motion. The specific problem sequences to be included in the instruction will be determined by our analyses of which topics are necessary to induce changes in students' schemata and by determining how far along the continuum of approximating the physics experts' schemata we wish the students to proceed.

For assessing the students' motion-of-objects schemata before and after instruction, we shall use several different techniques drawn from our own and other researchers' studies. Each technique probes students' declarative knowledge in a somewhat different fashion, yielding unique shards of data for analysis. By combining the results of the analyses of the data from several sources, we should obtain rather complete descriptions of the contents and structure of students' schemata. The assessment techniques we plan to use include DOE tasks, which we have used in previous studies (e.g., Champagne, Klopfer, & Anderson, 1980; Champagne, Klopfer, Solomon, & Cahn, 1980), and the interview-about-instances technique (Osborne & Gilbert, 1980).

Asking subjects to sort physics problems into types according to their own criteria and obtaining think-aloud protocols as subjects solve problems are two techniques used to study the representation of physics problems by experts and novices (Chi et al., 1981). We shall apply these techniques in our student assessments, except we shall restrict the domain to the qualitative mechanics problems in which we are interested. From analyses of the students' performances and explanations, we shall learn whether the students attend to surface features or to second-order features in analyzing problems, and we shall be able to infer whether the students' schemata are oriented by the objects of the problem or by a physical principle. To monitor the students' schema changes as they engage in instructional dialogues, techniques of protocol analysis, as suggested for example, by Ericksson & Simon (1980), will be applied. This research will provide significant empirical tests of the proposed instructional strategies and the psychological theory on which they are based.

REFERENCES

Anderson, J. (1983). *Architecture of cognition.* Cambridge, MA: Harvard University Press.
Anderson, R. C. (1977). The notion of schemata and the educational enterprise: General discussion of the conference. In R. C. Anderson, R. J. Spiro, & W. E. Montague (Eds.), *Schooling and the acquisition of knowledge.* Hillsdale, NJ: Erlbaum.
Anderson, R. C., Reynolds, R. E., Schallert, D. L., & Goetz, E. T. (1977). Frameworks for comprehending discourse. *American Educational Research Journal, 14,* 367–381.
Arons, A. B. (1976). Cultivating the capacity for formal reasoning: Objectives and procedures in an introductory physical science course. *American Journal of Physics, 44*(9), 834–838.
Atkinson, E. P. (1980). *Instruction-memory-performance.* Unpublished doctoral dissertation, Monash University, Melbourne, Australia.
Bates, G. (1978). The role of the laboratory in secondary school science programs. In M. B. Rowe (Ed.), *What research says to the science teacher.* Washington, DC: National Science Teachers Association.
Bransford, J. D., & McCarrell, N. S. (1974). A sketch of a cognitive approach to comprehension. In W. R. Weimer and D. S. Palermo (Eds.), *Cognition and the symbolic processes.* Hillsdale, NJ: Erlbaum.
Champagne, A. B., Gunstone, R. F., & Klopfer, L. E. (1982, May). *A perspective on the differences between expert and novice performance in solving physics problems.* Paper given at the annual meeting of the Australian Science Education Research Association, Sydney.
Champagne, A. B., Klopfer, L. E., & Anderson, J. H. (1980). Factors influencing the learning of classical mechanics. *American Journal of Physics, 48,* 1074–1079.
Champagne, A. B., Klopfer, L. E., & Gunstone, R. F. (1982). Cognitive research and the design of science instruction. *Educational Psychologist, 17,* 31–53.

Champagne, A. B., Klopfer, L. E., Solomon, C. A., & Cahn, A. D. (1980). *Interactions of students' knowledge with their comprehension and design of science experiments.* University of Pittsburgh, Learning Research and Development Center Publication Series, 1980/9. (ERIC Document Reproduction Service No. ED 188 950)

Chi, M. T. H., Feltovich, P. J., & Glaser, R. (1981). Categorization and representation of physics problems by experts and novices. *Cognitive Science, 5,* 121–152.

Collins, A. (1977). Processes in acquiring knowledge. In R. C. Anderson, R. J. Spiro, & W. C. Montague (Eds.), *Schooling and the acquisition of knowledge.* Hillsdale, NJ: Erlbaum.

Collins, A., & Stevens, A. L. (1981). Goals and strategies of inquiry teachers. In R. Glaser (Ed.), *Advances in instructional psychology* (Vol. II). Hillsdale, NJ: Erlbaum.

Dijksterhuis, E. J. (1969). *The mechanization of the world picture.* London: Oxford University Press.

Driver, R. (1973). *The representation of conceptual frameworks in young adolescent science students.* Unpublished doctoral dissertation, University of Illinois.

Driver, R., & Easley, J. (1978). Pupils and paradigms: A review of literature related to concept development in adolescent science students. *Studies in Science Education, 5,* 61–84.

Duran, R. P. (1980, October). *Influences of practical and instrumental knowledge on problem solving.* Paper presented at the NIE–LRDC Conference on Thinking and Learning Skills, Pittsburgh, PA.

Ericsson, K. A., & Simon, H. A. (1980). Verbal reports as data. *Psychological Review, 87,* 215–251.

Gagné, R. M., & White, R. T. (1978). Memory structures and learning outcome. *Review of Educational Research, 48,* 187–222.

Gee, B. (1978). Models as a pedagogical tool: Can we learn from Maxwell? *Physics Education, 13,* 287–291.

Gentner, D. (1980). *The structure of analogical models in science* (BBN Report No. 4451). Cambridge, MA: Bolt, Beranek & Newman.

Gerson, R., & Primrose, R. (1977). Results of a remedial laboratory program based on a Piaget model for engineering and science freshmen. *American Journal of Physics, 45,* 649–651.

Gunstone, R. F. (1980). *Structural outcomes of physics instruction.* Unpublished doctoral dissertation, Monash University, Melbourne, Australia.

Gunstone, R. F., & White, R. T. (1981). Understanding of gravity. *Science Education, 65,* 291–300.

Halley, J. W., & Eaton, B. (1975). A course in physics of human motion. *American Journal of Physics, 43,* 1007–1010.

Harre, R. (1965). *An introduction to the logic of the sciences.* London: Macmillan.

Hudson, H. T., & McIntire, W. R. (1977). Correlation between mathematical skills and the success of physics. *American Journal of Physics, 45,* 470–471.

Kolody, G. O. (1977). How students learn velocity and acceleration. *Journal of College Science Teaching, 6,* 224–227.

Leboutet-Barrell, L. (1976). Concepts of mechanics in young people. *Physics Education, 11,* 462–466.

Lindsay, P. H., & Norman, D. (1972). *An introduction to psychology.* New York: Academic Press.

Mallinson, G. G. (1977). Summary of research in science education–1975. *Science Education,* special issue.

Nagel, E. (1961). *The structure of science: Problems in the logic of scientific explanation.* London: Routledge & Kegan Paul.

Neisser, U. (1976). General, academic, and artificial intelligence. In L. B. Resnick (Ed.), *The nature of intelligence.* Hillsdale, NJ: Erlbaum.

Osborne, R. J., & Gilbert, J. K. (1980). A technique for exploring students' views of the world. *Physics Education, 15,* 376–379.

Peterson, K. D. (1979). The albertus magnus lyceum: A thomistic approach to science education. *Dissertation Abstracts International, 38*(1), 151–A.

Renner, J. W., Grant, R. M., & Sutherland, J. (1978). Content and concrete thought. *Science Education, 62,* 215–221.

Scheffler, I. (1969). *The anatomy of inquiry: Philosophical studies in the theory of science.* New York: Knopf.

Singer, B., & Benassi, V. A. (1981). Occult beliefs: Media distortions, social uncertainty, and deficiencies of human reasoning seem to be at the basis of occult beliefs. *American Scientist, 69,* 49–55.

Weller, C. M. (1970). The role of analogy in teaching science. *Journal of Research in Science Teaching, 7,* 113–119.

West, L. H. T. (1980). Towards descriptions of the cognitive structures of science students. In W. F. Archenhold, R. H. Driver, A. Orton, & C. Wood-Robinson (Eds.), *Cognitive development research in science and mathematics.* Leeds: University of Leeds.

White, R. T. (1979). Relevance of practical work to comprehension of physics. *Physics Education, 14,* 384–387.

White, R. T., & Gunstone, R. F. (1980, November). *Converting memory protocols to scores on several dimensions.* Paper presented at the annual meeting of the Australian Association for Research in Education, Sydney, Australia.

6

LANGUAGE, UNDERSTANDING, AND COMMITMENT

John O. Head and Clive R. Sutton

One of the weaknesses of the "alternative framework" description of learners' thought is its vagueness about change. Why should one person's pattern of understanding be resistant to change, and another's less so? What can a teacher do about it? We believe that openness to change is a function of feeling, and that how you feel is linked with how you think and how you talk. Cognitive, affective factors, and choice of interpretive language are interrelated. In this chapter, we develop a model of their interdependence within a person's "mosaic" of thought. We suggest that some areas of coherence produce commitments which contribute to a one's sense of personal identity. Change occurs readily in relation to the strengthening of that identity.

The model has been developed on the basis of assumptions which are not likely to be controversial within the context of this book. Nevertheless, to make them explicit will clarify our reasoning.

SOME BASIC ASSUMPTIONS ABOUT COGNITION, LANGUAGE, AND MOTIVATION

A Major Motivating Force for Human Beings is Their Need to Make Sense of Their World. This belief is, of course, the core of Kelly's (1955) personal construct theory. This is compatible with the evidence that people are naturally mentally active as evidenced, for example, in the findings from sleep research and from studies of sensory deprivation. We believe such mental activity to be important, as it gives a different perspective to the issue of motivation. Most theories in the past assumed that essentially people were inert until prodded or moved to be otherwise. Hence motivation was commonly described in terms of needs and drives. If, however, we assume that people are naturally active, then studies of motivation are not concerned about why a person chooses to do something rather than nothing, but why that person chooses one specific activity in preference to the range of possible alternatives. Hence the question of motivation is framed in terms of factors influencing the making of that choice. We see these choices as related to how the person concerned is attempting to extend fragments of his or her cognitive structure in the effort to make sense of experience.

Most of Our Mental Constructs Have a Strong Language Component. The resolution of a problem in understanding things usually involves adopting a particular way of talking about them. If someone is struggling to make sense of why this pan handle burns while that one does not, they may arrive at some point of rest—at least for the time being—by talking about heat "flowing" or being "conducted" along the metal. As it happens, these particular words and their associated imagery have been shown to be not wholly satisfactory to account for all the phenomena of heat. Nevertheless they still serve many people, helping them to make an island of sense in a sea of otherwise disordered impressions. Words can form centers for the crystallization of ideas.

THE SENSE MAKING EXPERIENCE IS AN IMPORTANT
SOURCE OF EMOTIONAL SATISFACTION

Once one gets "inside" a theory, it starts to seem "right." The subjective experience is one of insight and it engenders a feeling of truth—the insight rapidly becomes an outlook. For example, once you have come to the view that the state of the nation is accounted for by the existence of 'Reds under the beds' all further information will be processed accordingly. Already at this point, we begin to see how commitment grows out of particular cognitions, and in turn will shape further development of the cognitive structures.

COGNITIVE STRUCTURES AS MOSAICS

With these points in mind, how should we describe the complex of thoughts and ideas that a person has? We now attempt a description of its organization in terms of small areas of coherence, discrete and separate at first, capable of being isolated from each other; areas of mismatch and conflict, but also, in some cases, capable of being patterned together into a larger whole.

Structures are Built Up from Discrete Parts. We have many different experiences that we make sense of gradually, a bit here and a bit there as we meet them. On some occasions, links can be made between these bits so that a wider sense is made, incorporating previously separated understandings (superordinate learning, in Ausubel's terms).

There are many images that could be used to explore this idea of cognitive structures and we prefer the term *mosaic* to the alternatives in the literature. This particular image suggests that

1. each person idiosyncratically builds his own mosaic, and
2. the tiles employed by the individual are limited in range both by the constraints of language and of personal experience. Just as some Roman pavements were made from fragments of Greek temples, so a person can rework previous ideas to generate new patterns. The making of a mosaic is constrained by the color and the size of available tiles; similarly, the availability of concepts is partially determined by the semantic history of the language in which they are framed.

At a certain stage in the development of a mosaic, the patterning on a large scale becomes more obvious and important than the components, and areas of coherence can be discerned, giving an individual identity to the particular mosaic.

From the above, it is clear that we see human beings as pattern-seeking organisms, and where they achieve success in patterning they are likely to be motivated to continue, but only in accordance with the merging pattern. Gradually, the accumulated cognitions build up to form, in effect, the individuality and identity of their possessor.

Of course, we recognize that this mosaic image, like all analogies, has limitations. It does not deal adequately with the fluidity of thought, nor is a two-dimensional picture as good in some respects as one which can represent multidimentional connections. All analogies should be inspected for such points of mismatch to avoid an unwarranted reification of the model. Nevertheless, the mosaic analogy has further heuristic uses, as follows:

A Cognitive Structure Conceived in This Way Might Contain Structural Defects, or Points of Dissonance. These could be of two kinds:

1. *Internal dissonance,* in which separate parts of the mosaic cannot cohere, that is, the person holds simultaneously mutually contradictory views.
2. *External dissonance,* in which the constructs held already are incompatible with the input of new information.

The presence of dissonance might act as a stimulus to mental activity and the development of new concepts and structures. On the other hand, dissonance can be avoided by ignoring the input or by employing dismissive statements. Dismissive statements remove the need to examine the issues involved. They may account for the available information but are not open to further development. For example, the belief that "I cannot understand science" removes the need to try. Although the excessive use of dismissive statements is characteristic of closed minds, probably all persons need to make some use of them in order to avoid cognitive overload and to allow the concentration of attention on issues seen to be crucial.

The Set of Constructs That Relate to Oneself and One's Preferred Activities ("Things I Know I'm Good At") are a Particularly Important Part of the Mosaic. Making sense of electric circuits is just one small part of life, but making sense of oneself and one's relationships to other people permeates very large sections of experience. If the electric circuit activity also contributes to the latter problem—"I'm a person who understands physics"—it can be doubly effective in lending coherence to the mosaic.

One's overall sense of identity, of difference from other people, and of similarity to members of one's reference groups, arises out of the patterning that goes on in understanding all aspects of the world. The patterns one builds up are distinctive but related to those of other people, and this is one of the ways in which there is interaction between what have commonly been regarded as separate cognitive and affective factors. The need to make sense of the world (Assumption 1) is essentially an emotional condition whether its immediate object is overtly personal (i.e., "Who am I?") or more indirectly so (i.e., "This circuitry stuff . . ."). In a way, there is no knowledge that is not personal. Moreover, as cognitive beliefs represent models of the world and cannot be established as being true in an absolute sense, the choice of one cognitive view in preference to alternatives is likely to be made partially on subjective grounds. That statement does not deny the possibility that there may be sound objective reasons for preferring one theory in, say, physics to another (e.g., one theory may appear to fit the data more closely, but there are no objective grounds for stating that a belief is absolutely and always true).

Also, belief systems tend to carry hidden implicit metaphors, often possessing a strong emotional quality, so the acceptance of a belief system carries with it acceptance of these hidden values as well as the more objectively stated, explicit beliefs. A girl deciding to specialize in physics, and thus labelling herself a "physicist," not only accepts the models and paradigms of formal physics teaching, but has also chosen to join an identifiable group of the population with a distinct image and distinct group norms. She has chosen to enter a group which is predominantly male, which may possess a stereotypical image (of the sober, reticent controlled scientist), and which operates according to a set of definitive norms. In that event, the choice involves much more than an ability and interest in physics as such.

FROM UNDERSTANDING TO COMMITMENT

There are a number of possible reactions a person can give to a newly encountered idea. As mentioned above, it might be ignored or relegated to a low status by a dismissive statement. A full understanding often cannot be achieved quickly, so if the idea is grasped at all, it will usually be through rote learning. The idea is then no part of the individual's repertoire of general sense-making beliefs about the world; it will merely be a context-bound formula to apply in a given situation. Many of us probably learned differential calculus in that fashion, being able to carry out the necessary manipulations without understanding what the process really was. Only later we may, if we are lucky, acquire some understanding of what we have been doing when carrying out the required procedures.

Once the new concept has been successfully integrated, or in Ausubel's (1968) terms "subsumed," into the individual's existing, personal cognitive structure, it becomes part of that person's repertoire of tools used to make sense of the world. At that stage, the individual readily acquires an emotional attachment to the idea; the result is a commitment to a belief. That commitment helps define the identity of the person. We can be described in terms of what we believe. Certainly psychologists do so in talking of an authoritarian personality (Adorno, Frenkel-Brunswick, Levinson, and Sanford, 1950) who might be recognized by the possession of a particular set of attitudes and values.

Commitment is a double-edged sword. On the one hand, we can appreciate that unless a person has a firm set of beliefs and values he or she is likely to suffer from ego diffusion, resulting in ineffective behavior. There is, however, the other danger that a person becomes committed to beliefs which are incorrect, unhelpful, or damaging, and might resist attempts to produce change. We can recognize too, a qualitative difference between the possession of firm beliefs accompanied by a measure of flexibility and a willingness to consider alternatives, and the closed mind deaf to debate. (Rokeach, 1960) How might such qualitative differences arise?

The most likely explanation lies in the process of the acquisition of commitment. Mature choice involves reaching a firm decision after giving full thought to the issue. That process of thought, which might involve self-criticism and review of personal beliefs, can be painful; so a person might foreclose on a decision (i.e., make a firm decision without adequate thought) and thus, gain the sense of security and identity that the possession of the belief brings. In this case, further challenge to the belief will be resisted as it reopens the issues which had previously been deliberately brushed aside.

Such a person will tend to resist some aspects of learning which involve changing one's mind to committed ideas. This model of the dynamics of commitment has both general implications and a specific relevance to science education research. The general point is that as many issues cross the cognitive–affective boundary, it might be wise to make use of methods of inquiry and explanatory models drawn from both sides of that artificial division. Often the insights gained from one field can illuminate the other area. The specific relevance can be demonstrated in two instances: that relating personality and subject choice is spelled out in the next section; the other, that of the learner's often reported resistance to new ideas because of an apparent commitment to prior beliefs, also merits some attention.

Perhaps the main rationale for the study of cognitive structures is that it ought eventually yield useful pedagogic advice for the classroom teacher. There is now overwhelming evidence that many learners cling to their prior conceptions tenaciously in face of conflicting evidence and attempted persuasion; and, clearly, a better understanding of the dynamics of change, the requirements of fluidity of thought, are crucial. There have been suggestions in the recent literature, for example, by Hewson (1981) and Posner, Strike, Hewson, and Gertzog (1982), of the logical conditions necessary for an individual to be willing to change his or her mind and thus learn new ideas that conflict with existent beliefs. The main omission from such accounts has been an adequate recognition of the affective elements: changing one's mind is not a purely cognitive act! (West and Pines, 1982). The nature of commitment and the association of specific personality characteristics with authoritarian or closed minds are likely to be useful starting points for such studies, which, in turn, may be the most important work related to cognitive structures in the next few years.

SOME POSSIBLY USEFUL TECHNIQUES

SENTENCE COMPLETION TESTS

As part of a research study on personality and subject choice among high school students one of the authors (see Head & Shayer, 1980) made use of Loevinger's (1970) sentence completion tests of ego development. These yielded not only the information being sought to test a specific hypothesis, that boys opting for science might be less mature than those opting for the humanities, but also seemed to reveal further differences between subgroups of the population.

A process of item analysis revealed clear sex differences, over and above those of ego level scores. For example, the stem "I feel proud . . . ." produced more references to personal achievement from the boys and more references to receiving praise or compliments from the girls. With items involving interpersonal relationship, between the respondent and parent, for example, the boys were more selfish and exploitive, seeing parents in terms of what they provided, while girls placed more emphasis on the complex, changing, and reciprocal nature of the relationships. This finding that adolescent girls are more mature with respect to interpersonal relationships while boys are more concerned with their own future and achievement tells us nothing new; similar descriptions can be found in the literature (e.g., Douvan & Adelson, 1966), but it does illustrate the power of sentence completion tests in eliciting information.

Further differences were found between boys opting for science and those who chose humanities. The former group were emotionally reticent, demonstrating a reluctance to admit fears and anxieties, and displayed little sympathy for less fortunate persons. The stem "When a child will not join in group activities . . ." produced sensitive and tactful responses from the girls and boys opting for the humanities, suggesting that one should not force the child into anything, one should try to ascertain reasons, or one might offer some encouragement and help, and so forth. The boys showing a science preference often gave very unsympathetic responses such as "He is selfish," "He deserves to be alone," and "He should be made to conform." These differences between students opting for different subject specialties were often highly significant statistically, suggesting that interest in a subject is indicative of a total perspective or cognitive structure rather than just an ability to handle certain cognitive tasks.

Head (1980) suggested that these differences between students choosing different subject specialties could best be understood in terms of Marcia's (1966, 1976) descriptions of ego identity statuses. Within our prevailing culture, boys who are in the foreclosure state might be attracted to science, as it is emotionally undemanding yet is potentially useful, whereas boys who are experiencing moratorium will probably find the instrumental values of science scarcely relevant to their current preoccupations. For girls to choose science, which goes against the general trend, probably requires both commitment and self-examination ('crisis' in Marcia's terms) and they will have already achieved a firm identity.

Sentence completion tests have long been used in personality research and allied fields, but their use to explore cognitive structures is less common. For that purpose, they offer two distinct virtues: they seem to be highly effective in eliciting full and frank responses from the subjects, and they allow the respondents to speak in their own words so that there is the freedom to say what they want to, what is in their minds. Many test procedures, for example, the multiple choice style, deny that freedom. Other procedures which provide such opportunity present other problems. For example, interviews can be time-consuming and subject to bias. Sentence completion tests seem to gain worthwhile responses from a wide ability range and can be relatively easily assessed. The fact that they yield information which would usually be described as relating to the affective as well as the cognitive is in our opinion a further strength in attempting to gain a fuller understanding of cognitive structures.

"IT'S AS IF . . ."

The partial freedom offered by the sentence completion test can be taken a step further, when children are faced with novel experiences and are completely free to select their own words. Inevitably, they draw on their previous experience and vocabulary. What they say not only gives us clues to the idiosyncratic experience of the novel situation by the child, but it can also provide us with insights into an important part of new thought—the beginning of an act of commitment to a way of looking at the situation: "It's a sort of gas candle," says an eleven-year-old, looking at a Bunsen burner. "It's like a lot of glowworms," says another, looking down a microscope at particles in Brownian motion. In the first case, the interpretation is highly appropriate—a new part of the child's cognitive structure that a science teacher might well encourage. To see a burner as a kind of candle opens up possibilities for a unified view of combustion in the way that we (the scientific community) have seen it ever since Lavoisier's time. In the other example, points of positive analogy with glowworms undoubtedly exist, but this particular description misses the essential passivity of the Brownian particles (as Brown "saw" them), subjected to random buffeting by unseen missiles. Both, however, are *experiments in ways of seeing and ways of talking about* experiences.

To understand the growth points of a learner's cognitive structure, we suggest that greater attention should be paid to the child's experience and expression. The case for doing so comes partly from the history of science, in which the 'seeing as' phenomenon has played such a prominent part, initiating the development of new structures of thought. The shift from seeing combustion as the escape of phlogiston to seeing it as an uptake of something from the air is just one example, and such shifts are invariably accompanied by changes in preference for certain vocabulary.

Those who first talked about "harnessing" water power, and of measuring the 'horsepower' of engines, drew these words easily from the common experience of the times. However, these words have lost the quality of live metaphors, and become literalized—even technical terms—as have *charges* of electricity, *capacity*, and *current*. People, as it were, "buy" the metaphor, take on the new way of thinking, and start to take it for granted.

This phenomenon (also known as the reification of constructs), is now widely recognized as important in the intellectual history of individuals and the growth of public knowledge. One of us (Sutton, 1978, 1980a, 1980b, 1981a, 1981b) has already described it in terms of *insight* (gained from a new live metaphor), *doubt*, and *devotion* (as the insight becomes an indispensible part of the person's world view). Cognition, emotion, and rhetoric are not at all separate in this phenomenon, as words persuade people into new cognitions. The point we wish to make in the context of this book is that it is important enough to justify a search for children's spontaneous articulations of their insights, and to provide positive encouragement for them to do so.

CONCLUSIONS

What is involved in changing your mind?

As learners encounter new experience, they face the possibility of changing their minds all the time, sometimes by little adjustments, sometimes by major shifts. They may also close off and resist change, because changing one's mind is more than just a cognitive matter. To investigate this phenomenon properly, techniques are required that do justice to its personal significance and its emotional importance to the learner. In a small way, the two techniques described are a start in that direction.

In eliciting information about cognitive structures, we suggest the inclusion of the following techniques alongside others already available:

1. sentence completion tests as indicators of the affective aspects of how learners see themselves;
2. other measures of self-concept;
3. encouragement of children to articulate their insights in metaphorical or analogical terms, and the collection of their spontaneous productions of this type.

In representing cognitive structures, a form of notation is needed which

1. takes into account the growth of separate and discrete parts in which we have called the mosaic of thought;
2. highlights structural defects;
3. indicates how key areas of a person's cognitive structure contribute to his or her sense of identity.

REFERENCES

Adorno, T. W., Frenkel-Brunswick, E., Levinson, D. J., and Sanford, R. N. (1950). *The authoritarian personality.* New York: Harper.

Ausubel, D. P. (1968). *Educational psychology: A cognitive view.* New York: Holt, Rinhart, and Winston.

Douvan, E. & Adelson, J. (1966). *The adolescent experience.* New York: Wiley.

Head, J. (1980). A model to link personality characteristics to a preference for science. *European Journal of Science Education, 2,* 295–300.

Head, J. & Shayer, M. (1980). Loevinger's Ego Development Measures—A new research tool? *British Educational Research Journal, 6,* 21–27.

Hewson, P. W. (1981). A conceptual change approach to learning science. *European Journal of Science Education, 3,* 383–396.

Kelly, G. A. (1955). *The psychology of personal constructs.* New York: Norton.

Loevinger, J. & Wessler, R. (1970). *Measuring ego development 1,* San Francisco: Jossey-Bass.

Loevinger, J. Wessler, R., & Redmore, C. (1970). *Measuring ego development 2.* San Francisco: Jossey-Bass.

Marcia, J. E. (1966). Development and validation of ego-development status. *Journal of Personality and Social Psychology, 3,* 551–558.

Marcia, J. E. (1976). *Studies in ego identity.* Unpublished manuscript.

Posner, G. J., Strike, K. A., Hewson, P. W., & Gertzog, W. A. (1982). Accommodation of a scientific conception: Towards a theory of conceptual change. *Science Education, 66,* 211–227.

Rokeach, M. (1960). *The open and closed mind: Investigations into the nature of belief systems and personality systems.* New York: Basic Books.

Sutton, C. R. (1978). *Metaphorically speaking: The role of metaphor in teaching and learning science.* Occasional Papers, Leicester University School of Education.

Sutton, C. R. (1980a). The learner's prior knowledge. *European Journal of Science Education, 2,* 107–120.

Sutton, C. R. (1980b). Science language and meaning. *School Science Review, 62,* 47–56.

Sutton, C. R. (1981a). (Ed.) *Communicating in the classroom.* London: Hodder and Stoughton.

Sutton, C. R. (1981b, Spring). Metaphorical imagery: A means of coping with complex and unfamiliar information in science. *Durham and Newcastle Research Review, 9,* 216–222.

West, L. H. T., and Pines, A. L. (1982). How 'rational' is rationality? *Science Education, 61,* 37–39.

7

TOWARD A TAXONOMY OF CONCEPTUAL RELATIONS AND THE IMPLICATIONS FOR THE EVALUATION OF COGNITIVE STRUCTURES

A. Leon Pines

What is the meaning of *cognitive structure?* The words give us important clues. *Cognitive* means "of the mind, having the power to know, recognize, and conceive, concerning personally acquired knowledge," so cognitive structure concerns individual's ideas, meanings, concepts, cognitions, and so on. *Structure* refers to the form, the arrangement of elements or parts of anything, the manner or organization; the emphasis here is not on the elements, although they are important to a structure, but on the way those elements are bound together.

These two components of cognitive structure are the components in the analyses presented in this chapter. What binds them together are "relations." Meaning, thought, concepts, and language depend on relations. So too does their structure. The meaning that an individual gives to a particular word, and the complex conceptual framework that an individual possesses which makes him or her knowledgeable in a particular area both depend significantly on relations.

Consequently, in this chapter, I explore the nature of relations, interpret meaning, thought, language, and concepts from the perspective of relations, and begin the process of developing a taxonomy of relations. Although this latter task is not complete, I indicate some implications of this approach for future research on the evaluation of cognitive structures and for the improvement of curriculum and instruction.

RELATIONS

Relations exist both between objects and events in the world and between concepts and propositions that denote these objects and events. We use symbols such as words (or any other notational system) to express, manipulate, and communicate these relations. All meaning is relational! If this is true, and I will endeavour to demonstrate that it is, then it follows that the elucidation of meaning is only possible through the analysis of relations.

Complexity—in the real world and of meanings—is a result not of the quantity of relations among elements, but of a combination of both quantity and quality of relations, that is, the number and types of relations that exist. A multiplicity of relations among identical elements does not increase the complexity of meaning. For example, a small pile of sand or salt is no more complex than a larger pile of sand or salt; and, the echolalia of a baby or autistic child is meaningless no matter how long the repetition continues. Once we take the sand and build structures with it or use the salt for different purposes or combine words into non-echolalic, non-repetitive sentences, then we begin to increase the complexity and potential complexity of meaning. In short then, complexity arises out of various elements related in various ways or, to put it another way, in order to exhibit complexity, elements and relations must vary. In analyzing complex structures, situations, and meanings, what will be revealed at the micro-level are different elements related in a variety of ways. Concepts, we will see, are complex summaries of numerous such relations that can be expressed as propositions.

MEANING, THOUGHT, AND LANGUAGE

Meanings are created by and within complex nervous systems. This is an important point. Meanings do not reside in the world, only potential meanings do! Before the dawn of evolution, when no organisms were alive, there were no meanings in the world. If all life should disappear from our world—a horrid but not unrealistic prospect—then again there would be no meanings in the world. As we move up the phylogenetic scale and nervous systems become more complex, there is the realization of more meaning. In human beings, *homo sapiens*, meanings can be created at a very abstract, symbolic level. This is unique to our species as a consequence of our large neo-cortex and our facility with language and other symbolic systems. However, although we are most concerned with human beings and complex, abstract learning and cognition, it must be emphasized that any organism with a complex enough nervous system is capable of extracting meanings from the world in which it lives. One way of saying this is that meanings are a consequence of the neo-cortex in action. The senses, along with the organizing power of the neo-cortex, enable the interpretation of sensations which are rendered in the perception of objects and events. Now clearly, the more developed the neo-cortex and the more learning that has taken place, the more discriminating and the more meaningful are these perceptions.

Of what importance is it to go further than this in elaborating the meaning of "meaning" and showing that all meanings are basically relations? Will this quasi-philosophical inquiry have anything to say to education? Part of this answer is well articulated by Macnamara (1982):

> There is enormous difficulty in specifying what meaning and reference are. Still nothing is to be gained by fighting shy of the difficult problems . . . If anyone can improve on the theories, he will, in doing so, have means for improving psychology. (p. 6)

Indeed, if we can understand meaning and, by doing so, improve our understanding of meaningful learning, then we will be in a position to improve education.

When an organism perceives, it is organizing sensations into meaningful patterns of relations. Certain things, so to speak, "go together." Either they are spatially organized into objects or temporally organized into events. The Humean principles of contiguity and cause and effect are based on such empiricist notions. The relations based upon proximity in time and space are picked up by human and subhuman species in the form of object permanance and causality. The infant establishes such relations within the first two years of life (Piaget, 1954). Whole behaviorist theories of learning are based on these notions of association and conditioning. These theories are not wrong, but they are incomplete. They can be readily applied to most subhuman species or to profoundly retarded human beings, but they are inadequate to explain meaningful learning in human beings who have a symbolic, abstract language with which they can think and communicate. The ability to mentally manipulate and communicate concepts adds a whole new dimension that is not evident in most, if not all, subhuman species.

Sensation—the raw data from the sense organs—on its own, without perceptual organization, is devoid of meaning. Organized sensation—namely, perception—enables the awareness and mental recording of objects and events. In human beings, such perception is facilitated by language—words or sentences, and thus experience is conceptually and propositionally punctuated into meaningful distinctions, relations, and complexes of such relations that transform "raw sensation" into perception (i.e., meaningfully organized sensation). Again, the perception of objects and events is not unique to human beings, but evident in subhuman species too; what is unique to us however, is the abstract conceptualization and the use of language. In language, relations and whole networks of relations are frozen into concepts labeled by words. These concepts and words capture the way a culture slices up reality: what relations are considered important enough to discriminate and preserve, what patterns and regularities have been worthwhile preserving, and so forth. Such an evolution of language that serves to preserve conceptual structures and transmit them from generation to generation is only evident in the human species. This is what predominately distinguishes us from subhuman species and one good reason why behavioristic psychology is inadequate to describe or explain human thought, learning, and behavior.

Language serves the dual purpose of thought and communication. The ability to acquire and use language enables the amplification of meaningful experience. For one, it enables the transmission of culture from one generation to the next; for another, it enables individuals, in their ontogenetic development, not only to remember past events, but to acquire large bodies of information beyond their immediate experience, and to use this knowledge in their adaptation through prediction and problem solving. Not surprisingly then, the evolution of language capacity has brought about what seems to be a discontinuity in the phylogenetic scale. Biological evolution mitigates against the cultural evolution made possible by language. Education, and particularly schooling, is a means for transmitting extra biological meaning from generation to generation. A large part of this is already accomplished simply by virtue of language acquisition which incorporates within itself enormous amounts of conceptual content and cultural meaning.

I will discuss three types of meaning: signs, isomorphisms, and symbolic meaning. Relations, as I have already mentioned, are the basis of all meaning. Lest we get bogged down in a pure reductionism, I must, at this point, mention that no single meaningful relationship exists independently in the mind: All relations are embedded in a larger context, exist as part of a larger situation, matrix, or frame of reference. Thus we never find a simple diadic relationship of the A R B type (where A and B are elements, and R is relation); rather, we find A R B in the context of P or under conditions Q and S. Thus, in order to understand one single relationship, one needs to understand a whole context. Or, to cite Minsky (1982): "There never is much meaning until you join together many partial meanings; and if you have one, you haven't any" (p. 19).

Even so, we can, if pressed, find a simple relationship to make the point that all meaning is contingent on such basic relations. Thus, the "shore" or "coast" are meaningless without the relationship between sea and beach, or land.

With this said, we can momentarily return to the single unit of meaning. When we say that two objects or events have the relationship A R B, it is no different than saying that A means B (in the context of P or under conditions

Q and S). The most simple type of meaning is that of a sign. A is a sign of B. Our universe is made up of sign situations that are the basis of all meaning. The sound of barking is a sign for a dog, a wagging tail, a sign that there is a body and head attached, and the smell of smoke or the feeling of heat, that there is a fire. Signs[1] therefore are physical constituents, the necessary structure of the world, meaningful by virtue of our sensory apparatus. Such meanings are, of course, dependent upon the available sensory mechanisms and we could imagine different sign situations that might exist with different sensory capacities.

Often we can infer one thing from another based upon both sensory capacities and past experience. Thus, footprints on the beach are signs that someone has walked there, steam from a kettle is a sign that the water is boiling, and dark clouds are signs that it will rain. Using sophisticated detection instruments, we can extend our sensory capacities as with telescopes or radiotelescopes detecting such astronomical rarities as Sigmus XI which would otherwise be undetectable. We will return to this later. Interpreted signs of this sort can either lie in the existence of non-existence of phenomena, as when the heartbeat is a sign of life and the lack of a heartbeat often a sign of death. There is no reason to believe that subhuman species with sophisticated sensory apparatus are incapable of deriving such meanings from the world. For example, the sharp sense of smell in some animals enables them to detect odors that are signs of prey that they are stalking, or predators they must avoid. But these animals are not conscious of meanings as we human beings are. We may speak, therefore, of levels of meaning culminating in conscious meaning (which can itself be analyzed into different levels, depending upon the nature and extent of symbolic relations). Nevertheless, in all cases, the basic meaning is one of signs that exist by virtue of the nature of the world in which we live.

The second type of meaning that exists is that of isomorphism. When two things have a point to point correspondence with one another, so that each element of one corresponds to just one element of the other, and any relation between elements of one corresponds to just one relation between elements of the other, they are said to be isomorphic. Thus, one thing can mean another by virtue of an existing isomorphism. We may term such meaning "representational meaning." One thing represents another because there is a correspondence between them. Thus, projections have a sort of isomorphic relationship—a shadow being a representation of a body upon which a light shines. Photographs, paintings, and maps are also sorts of

[1]Sign is used differently in the linguistic, structuralist tradition of de Saussure (1966). His analysis is purely linguistic and refers to linguistic performance (parole) where the utterance is the signifer and the concept is the signified. For us, language is yet to come when we discuss symbolic relationships.

projections—isomorphic representations. However, isomorphisms are not limited to visual representations. We can code information in diverse ways. There is an isomorphism between the grooves in a record and the music which we hear when the record is played on an appropriate phonograph.[2] Thus, meaning can be captured and coded in isomorphic structures which can either be overtly observed as representations (e.g., photographs) or need to be decoded so that the meaning is extracted (e.g., phonograph records). Douglas Hofstadter's Pulitzer prize-winning book, *Gödel, Escher and Bach: An Eternal Golden Braid* (1979), is an analysis of isomorphisms. Hofstadter illustrates the isomorphisms inherent in mathematical and formal systems. According to him, all meaning is the detection of isomorphism. Information is meaningless without a code to detect isomorphisms. Different code systems can unravel different meanings from the same phenomena.

Much energy is spent by scientists and other individuals in making records that capture meaning. In art, painters and sculptors make records of objects, events, or emotions. Sometimes these records are clear, observable isomorphisms (naturalistic or realistic art); sometimes the records are more abstract and sometimes they necessitate interpretations or decoding (impressionism, expressionism, or abstract art). I do not wish to carry this analysis of art any further, because, clearly, isomorphism does not underlie all artistic creation; aesthetics is more than representational meaning. A better example is science. Here much is done to render the world meaningful in a less individually subjective way. Scientific theories are isomorphic with the particular aspect of reality that they model. Scientific measurements are isomorphisms. So too are logico-mathematical systems.

In science, attempts are made to capture events, to record them. Progress in science is facilitated by the invention of evermore sophisticated recording devices. These instruments extend human sensory capacities (such as telescopes and electron microscopes) or they enable the recording of phenomena otherwise inaccessible to our sensory mechanisms (such as certain long and short wave electromagnetic phenomena). The scientific world abounds with recording devices, such as oscilloscopes, seismographs, pH meters, and noise meters. Clearly these records undergo further interpretations. Data are organized and transformed after the recording, and conclusions are derived. The point is that just as meaning exists in sign situations, so too, meaning exists in projections or recordings that are isomorphic with these situations. Wherever a point for point isomorphism exists, there can be meaning. The representation by virtue of its correspondence contains meaning. The same information in the sign situation (or part of that informa-

[2]DNA is an excellent example of a complex, natural isomorphism. The chemical structure can be decoded rendering a functioning organism. The notion of a genetic code may be an appropriate metaphor because it is a means of understanding an isomorphic relationship.

tion) is contained in the isomorphic representation. Sometimes, when the pathway between the sign situation and record is convoluted, several sophisticated code systems may be necessary to go from one system to another isomorphic one. This is one reason why certain scientific or mathematical conclusions are meaningful to only a handful of experts who can decode them and follow the convoluted chains of isomorphisms.

This brings us to the third type of meaning, that is, symbolization. Unlike signs or isomorphisms, symbols are fragile, conventional meanings which are constructed by human beings. The most common symbols are words. These symbols are human inventions used to impose a sense of order on the world, and to enable communication. Although the way a culture slices up the world may not be totally arbitrary, the relation between words and related phenomena *is* conventional. The multiplicity of languages should suffice to make this point clear. Nevertheless, once a language exists, it imposes meaning on the world. Certain sign situations are captured by the language while others are ignored; hence the power of the Sapir-Whorf hypothesis that meaning is relative and culturally determined or dependent (Whorf 1956). Languages can and do change, and thus, over time, semantic growth or change is possible and evident. (One may even say that such change is inevitable.)[3]

We are not sure whether the deep structure of language (Chomsky 1957, 1972) is universal as a consequence of its biological origin. Surface structures and lexicons are obviously diverse and continually modified. The capacity for language is, to a large extent, what makes us human. Our ability to invent and use symbolic systems provides us with a qualitatively different experience—a far richer meaning—than that of subhuman species. The only meanings unique to human beings are those complex symbolic meanings as captured by our language. Simple signs or isomorphic representations (such as shadows) can be experienced by organisms with sophisticated nervous and sensory systems. Fish and bats, for example, create isomorphisms in their mediums that enable them to detect obstacles and to navigate (our radar and sonar systems are modeled on these types of isomorphisms). But only human beings derive meanings from such situations and experience by virtue of the symbols that literally modify the way signs and isomorphisms are sensed (we call this "perception"). Symbolization enables thought which in turn enables the construction of sophisticated projection and recording mechanisms that capture and preserve meaning ismorphically.

Finally, through the use of symbolic systems, non-existent frameworks of relations, which have no sign counterpart in the real world, can be created. Such human ontological creations are the basis not only for fiction, fantasy,

[3]Structural linguists often distinguish between the synchronic aspects of language, which exist at one particular time, and the diachronic aspects, which change.

and escape, but also are necessary in constructing scientific models and theories used to explain reality. Mythology and astrology are inventions made possible only because of an ability to create relations, to derive meaning from symbols; so too are poetry and science!

CONCEPTS

Conscious experience is mediated by concepts. Unlike signs and natural isomorphisms, concepts are cognitive entities; they are the furniture of the conscious mind. Conceptualization is another way of talking about conscious experience. The continuous flow of experience is punctuated, as it were, by the conceptual organization of such experience. Abstract conceptualization is made possible by the ability to symbolize.

Concepts are packages of meaning; they capture regularities (similarities and differences), patterns, or relationships among objects, events, and other concepts. Certain signs and groups are captured in referential conceptual relations; other concepts are invented at higher levels of abstraction by relating subordinate concepts to one another. Thus, the distinction exists between concrete concepts that denote referential relationships to middle size physical objects and temporally experienced events, and abstract concepts that portray relationships among other higher order concepts.

Each concrete concept is a human invention, a way of "slicing up" and organizing the world. Once labeled, they become communicable through the use of a language which, in turn, imposes organization on the world. Hence the power of an acquired language to channel human perception and thought; hence Wittgenstein's (1961) astute proclamation that the limits of his language were the limits of his world. But how much richer this symbolic world is over the evanescent reality experienced by non-conceptualizing organisms!

Concepts are regularities labeled with words and employed in thought and communication. The similarity between concepts and what I have previously termed *symbolic relations* should be evident. Conceptualization is indeed facilitated by the acquisition and use of symbolic relations. I would surmise that without an intricate symbolic system, conscious thought is greatly restricted and remains rudimentary, although pre-linguistic conceptualization is possible. This is true both for advanced primates and young children. In the mature linguistic adult, thought and language are inextricably intertwined (Vygotsky 1962) because of the fundamental role that symbols play in the manipulation of concepts in thought and communication. How strange that arbitrary conventions such as words can come to play such a fundamental role in conscious thought. A word is like a conceptual handle, enabling one to hold on to the concept and to manipulate it.

A concept is involved in an immense network of relations. A concept can be thought of as a theoretical point where meaningful relations converge, and each concept is a crossing point for a multitude of relations. These relations are the fibers from which meaning is constructed. No fiber exists in isolation. Bound with each other, they may form threads, or ropes, or complete fabrics of thought. The fibers are propositions, each signifying a single (or several) conceptual relation(s).

Concepts are neither true nor false. One can categorize concepts as more concrete or more abstract, more felicitous or less felicitous, more or less useful, and so on, but not as true, false, correct, or incorrect. Only when concepts are combined into propositions, do truth values arise. Propositions may be communicated by sentences. One might describe a specific concept as the hypothetical meeting place of all the propositional relations in which that concept participates. There are an infinite number of such relations, and a concept is a summary of all those relationships. A concept is a locus of meaning.

Whenever a concept has restricted meaning or can be summarized by a limited number of relations or properties, a simple definition in the form of an analytic proposition can be given. We can artificially restrict the meaning of a concept by stipulating definitions. This is often done in science. This gives the false notion that concepts are single units. By and large, however, concepts have various meanings within different contexts. In such cases, no context-independent, analytic-synthetic distinction can be drawn, and the notion of a concept as a category of reference which has a single meaning— an essence—is wrong. Rather, we find a concept as part of a semantic network, its meaning arising from a multitude of crisscrossing propositional relations. The analytic and synthetic dimensions dissolve into one another, depending upon how the concept is used.[4] For these reasons, although models of concepts as categories are tempting, and may even be useful, they are simply incorrect.

No single, best semantic network exists. Concepts may have a multitude of meanings and be used for a variety of purposes. Thus, a single concept may mean one thing within one framework and something slightly different within another context. Wittgenstein's (1953) notions of 'word games' and 'family resemblances' capture the diverse use of words or concepts from one context to another and the non-unitary meaning of words or concepts.

Because of the complexity of concepts, we should be willing to accept the fact that their acquisition is a long process which can never be complete. There is no such thing as the final acquisition of a concept. Rather, concept

[4]The analytic versus synthetic distinction in propositions is somewhat analogous to Rosch's (1973) distinction between categorical prototypes and border-line cases. A conceptual, prototypical relationship can be viewed as an analytic proposition. This analogy warrants further analysis but is beyond the scope of this chapter.

become differentiated in the mind of a person. As more and new relationships are acquired, the respective concepts take on new meaning. Some relations are, of course, more important than others. The only reason that different individuals can communicate and understand one another is because of the overlap between their conceptual-cognitive structures.

In short, concepts are never acquired in a finalistic fashion. Furthermore, certain conceptual relations that are acquired may be inappropriate within a certain context. We term such relations as "misconceptions." A misconception does not exist independently, but is contingent upon a certain existing conceptual framework. As conceptual frameworks change, what was deemed a misconception may no longer be a misconception; conversely, what is a central conceptual relationship in one framework may be a profound misconception within another framework. The history of science is replete with such examples.

At least two points must be made in summary. Firstly, concepts are not independent entities but, rather, are complex collections of relations embedded in a larger framework (or multiple frameworks). Secondly, these bundles of meaningful relations we call concepts are, on the one hand, capable of change, and, on the other hand, can never be acquired in any finalistic fashion. Any new relations will affect, to some extent, the total framework of relations.

Signs and simple isomorphisms are the essential furniture of the world. More complex isomorphisms and relevant decoding systems form bridges between relational systems; through the use of symbolization, signs and isomorphisms are conceptualized and become the furniture of the conscious mind. Signs, isomorphisms, and symbolic relations—which underlie all meaning—are the basis for human understanding and consciousness. Thus, relations (or patterns of relationships) are the basis for concepts, which, in turn, are the substance of thought.

Reality, in creatures with symbolic conceptual systems, is not perceived in an unbiased fashion. The world, figuratively speaking, is perceived by human beings through conceptual filters. Sensation— the raw data of the senses—is, in *homo sapiens*, transformed through conceptualization into perception. Thus, human meaning at all levels, from the perception of the world in which we live to higher abstract thought, is conceptual. Conscious meaning is achieved through conceptualization; concepts being those packages of meaning—systems of relations—conventionally labeled with words. Discerning or inventing a novel set of relations heretofore unknown in a culture calls for a new concept (or conceptual framework) which can only be grasped, even by its initiator, insofar as it can be articulated or communicated. The invention of a new word is not sufficient if that word cannot be explained and communicated in terms of other conceptual relations which already exist in the language. The implications are clear: two cultures may not be able to communicate beyond certain shared sign meanings, and, an individual deviant in a culture (one who has constructed alternative conceptual frameworks) will be understood only by those who can extend their own conceptual frameworks in an appropriate way.

Becoming a member of a society entails sharing meanings with other members of that society. In conscious organisms, these meanings are conceptual. In primitive nervous systems, no symbolization occurs; responses to stimuli or sign situations are acted upon reflexively, or, at best, habitually subsequent to conditioning; there is, however, no conscious meaning. In complex, sophisticated nervous systems, such as those of human beings, conceptualization emerges and with symbolization reaches new, heretofore impossible, levels of consciousness. Becoming a member of a human society entails the progressive acquisition of both concepts and language, each contingent upon, and influential of the other. Most human thought, therefore, is, finally, conceptual and mediated through language.

All domains of human knowledge are basically conceptual symbolic systems or frameworks. Clearly, knowledge can be expressed in action, but such knowledge can only be described or understood conceptually. Within a single culture, we may find subcultures that employ different conceptual frameworks and corresponding languages. Such highly specialized conceptual frameworks and languages cannot be understood by the uninitiated. What holds a culture together are shared systems of meaning; what may split cultures apart are unshared conceptual and symbolic systems. Highly specialized scientific disciplines, for example, are, or can become, subcultures of this sort. Until an individual becomes socialized, or initiated, so to speak, into a subculture—learning to comprehend its conceptual frameworks and corresponding languages—that individual will not be able to comprehend or communicate with members of that subculture on questions regarding the domain of interest of that subculture. Much of higher education and schooling in western civilization is the acquisition of specialized conceptual and symbolic systems, subsequent to the acquisition of more general conceptual and symbolic systems that bind a total culture together. These general and specific frameworks constitute the substance of nearly all early, elementary education and advanced education, respectively.

CONCEPTUAL RELATIONS

Thus far, we have briefly touched upon meaning and concepts, illustrating the central role of relations. The analysis is far from complete, for we have said nothing much of the distinction between information and knowledge

nor discussed what is to be considered as "understanding" or "knowing."
Epistemological analyses can be pushed further. All phenomena and records
of them are, for example, sources of information. Knowledge and understanding are the progressive conceptualization of such internalized information. Whereas information may exist in any natural phenomenon, it is meaningful only insofar as a nervous system can discern or create relations.
Knowledge is the process and product of conceptualization, the creation and
preservation of symbolic meaning. Understanding comprises meaningfully
acquired knowledge.

What is important at this point is the emphasis upon the primacy of
relations in all meaning; and, furthermore, the central role that symbolic
relations play in human cognition, in all human understanding. If relations
underlie all meaning, and concepts are the medium of thought and understanding, it follows then that we would greatly profit from an analysis of
conceptual relations. That is the thrust behind my proposal to construct a
taxonomy of conceptual relations. What types of conceptual relations exist?
Answering this question may provide a new, fruitful paradigm for research in
education.

Let me illustrate this point with a rudimentary analysis of conceptual
relations. I will refer to two fundamental types of conceptual relationships:

1.   Set–element relationships, (i.e., set–subset), and
2.   Whole–part relationships, (i.e., system–element).

Set—element relations underlie all classification systems. Classification
entails the grouping of anything, based upon certain properties, attributes,
or characteristics. Groups can be further classified leading to hierarchies or
taxonomies. In short the set–element relationship seems to be a basic synthetic organization which underlies much of human conceptualization, especially in Western civilization. Not only can individuals, objects, events,
groups, and so on, be classified, but the very attributes, properties, and
characteristics used to classify can themselves be classified. We refer to the
basic means of classification as qualification and quantification. That is,
groups can be formed on the basis of certain properties or on the basis of the
amount of a certain property or properties. It seems that in one swoop, we
have covered much of conceptualization with one single relationship: the
set–element.

Is it possible, therefore, that the set–element relationship can serve as the
basis for a taxonomy of relations, all other relationships being subordinated
to this relationship? Frege (1966) showed that the whole–part relation alone
would not serve as a fundamental relation but that element-hood might. But
this refers to logical rather than psychological matters. What is sufficient for
a logical structure may be insufficient for psychological structures. It is

possible to do arithmetic in set-theory, but as the failure of the New Math showed, that is not *how* we *do* it. Psychologically, whole–part relations may operate independently of set-element relations.

The whole–part relationship refers to the organization of structures and systems. For example, the structure of a body can be analyzed into its constituent parts; an event can be analyzed in terms of its stages. Thus, an analysis can always be seen in terms of a whole–part relation. A functional system can be dissected spatially or temporally. That is, we can always ask the question: What is X made out of? Or, if X is a functional system, we can ask: How does X work?

Again, the whole–part relationship seems to subsume much of what can be considered as human knowledge. Taken together, then, set-subset and whole–part relationships can serve as a basis for a taxonomy of relations. There may be other basic relations, but these two can for now suffice to illustrate the usefulness of pursuing such a mode of inquiry. Future analysis will focus on other types of relations. For example, causal relations can be, on the one hand, classified (set–subset) and, on the other hand, analyzed as temporal events (whole–part). Other complex relations such as metaphors must be carefully investigated.

IMPLICATIONS FOR THE EVALUATION OF COGNITIVE STRUCTURE

This represents only a first step toward a taxonomy of relations in content and cognitive structures. Its completion will not be the achievement of a single investigator, but the cumulative achievement of many able groups. In its current crude form, however, this analysis has many implications for the study of cognitive structure and, indeed, for the study of education in general.

Any study of the cognitive structure of an individual needs to be epistemological as well as psychological. I have already implied that, although meaning is personal, the coherence between personal meanings within a discipline is what constitutes that discipline. So the evaluation of the cognitive structures of individuals involves the evaluation of the content in a discipline or subdiscipline as well as the evaluation of the cognition of an individual. Both of these evaluations will be aided by a taxonomy of relations.

Let me consider content first. What types of relations exist in certain bodies of knowledge? For example, using the two types of relations that I have briefly analyzed in the previous section, we can clearly see that different aspects of biology rely more heavily on different relations. Taxonomic biology is based almost entirely on set–subset relations, grouping organisms into a hierarchical organization of kingdoms, phyla, classes, orders, families, genera, and species. Anatomy and physiology, in contrast, are more fruitfully analyzed as whole–part relations. Anatomy is structurally oriented, investigating the spatial organization or organisms, looking at how things are built, and breaking organisms into organs or structured systems, such as the skeleton into bones. Finally, physiology investigates how biological systems function, and how the anatomical components work in coordination to adapt, sustain, and perpetuate life.

Clearly these divisions of biology are gross and by no means mutually exclusive. But then, so too is our currently available taxonomy of conceptual relations. The way we classify organisms from a taxonomic point of view is much dependent upon certain anatomical or structural considerations; the ways organisms function physiologically is determined, to a large extent, by their structural components. Furthermore, certain sub-domains of biology have been invented to link biology to other domains, such as biophysics, biochemistry, sociobiology, and so on. And, within biology, we speak of areas such as neurophysiology, kidney physiology, the physiology of respiration, and so on, as links between anatomy and physiology; or mammalian physiology, the water physiology of the camel or the desert rat, as links between taxonomy, anatomy, and physiology. Moreover, whole biological subdisciplines such as genetics, at the macro-Mendelian level, or at the micro-molecular level, have their taxonomic, anatomical, and physiological aspects, as do botany, microbiology, and mycology. All this richness and complexity notwithstanding, we seem to be able to gain some insight into the nature of each of these complex areas, such as biology, by considering the nature of the prevalent conceptual relations in each sub-domain.

In my view, these basic conceptual relations exist in all disciplined knowledge. But a finer taxonomy of conceptual relations may help us to understand the similarities and differences in knowledge structures. Obviously, each discipline investigates different phenomena, but each deals too with conceptual relations; for, as we have seen, relations underlie all meaning.

And what of cognition? How are different relations acquired? Are some relations easier to acquire than others? Are there developmental trends or are some relations more basic than others? And, can we teach some types of relations more readily than others? A taxonomy of conceptual relations would also aid in the methodologies used for evaluation. The administration of clinical interviews to probe cognitive structure and to reveal existing relevant prior knowledge (Pines et al., 1978) might be more efficiently carried out. The task of pinpointing misconceptions would also be easier if we had a systematic organization and understanding of conceptual relations. For example, a misconception may be the result of using a set–subset relationship when a whole–part relationship is warranted.

Cognition deals with the acquisition, retention, and manipulation of meanings. These meanings are relational. Sensation, perception, and attention deal with the "extractions" of relations from our environments; memory is, by and large, the "storage" of meaningful relations and their use in thought and problem solving. Understanding the nature of relations, therefore, is a giant step toward understanding cognition.

CONCLUSION

In this chapter, I have attempted first to provide an analysis of meaning demonstrating that relations underlie all meaning. Secondly, concepts were shown to be the medium of thought, a symbolic network of relations through which organisms such as *homo sapiens*, with complex nervous systems, experience reality.

An attempt to investigate the basis of human understanding is certainly no new endeavor; it has been the realm of philosophers and psychologists for ages. Many have attempted to provide catalogues of categories to explain the entities of meaning or thought. Most have included within these catalogues the category of "relationship." I wish to subsume all meaning under this single category with the hope that it will provide a new avenue for fruitful educational investigation and practice. The example from biology is not meant to elevate biology as the candidate for such analyses, but merely to point out the value of such an approach.

Providing a taxonomy of conceptual relations, if it can be achieved, will, I predict, have profound ramifications for research in cognition, developmental psychology, and epistemology. All these, and other relevant areas too numerous to list, will influence curriculum development, instructional planning, teaching, and learning, which are the cornerstones of education.

REFERENCES

Chomsky, N. (1957). *Syntactic structures.* The Hague: Mouton.

Chomsky, N. (1972). *Language and mind.* New York: Harcourt Brace and Jovanovich.

de Saussure, F. (1966). *Course in general linguistics.* London: McGraw-Hill.

Frege, G. (1966). *A critical elucidation of some points in E. Schroeder's vorlesungen uber die algebra der logik.* In P. Geach, and M. Black, (Eds.) *Translations from the philosophical writings of Gottlob Frege.* Oxford: Blackwell.

Hofstadter, D. (1979). *Gödel, Escher, and Bach: An eternal golden braid.* New York: Basic Books.

Macnamara, J. (1982). *Names for things: A study of human learning.* Cambridge, MA: MIT Press.

Minsky, M. (1982). Why people think computers can't. *AI Magazine, 3*(4), 3–15.

Piaget, J. (1954). *The construction of reality in the child.* (Margaret Cook, Trans.). NY: Basic Books.

Pines, A. L., Novak, J. D., Posner, G. J., & Van Kirk, J. (1978). *The use of clinical interviews to assess cognitive structure.* (Curriculum and Instruction Series, Monograph #6), New York: Cornell University, Department of Education.

Rosch, E. (1973). Natural categories. *Cognitive Psychology, 4,* 328–350.

Vygotsky, L. S. (1962). *Thought and language* (E. Hanfman and G. Vakar, Trans.). Cambridge, MA: MIT Press.

Whorf, B. L. (1956). *Language, thought and reality: Selected writings of Benjamin Lee Whorf.* Cambridge, MA: MIT Press.

Wittgenstein, L. (1953). *Philosophical investigations.* New York: Macmillan.

Wittgenstein, L. (1961). *Tractatus logico-philosophicus.* London: Routledge and Kegan Paul.

KNOWLEDGE REPRESENTATION, COGNITIVE STRUCTURE, AND SCHOOL LEARNING: A HISTORICAL PERSPECTIVE

Thomas J. Shuell

INTRODUCTION

The chapters in Part I of this book represent a relatively new approach to an important and perennial concern—namely, the most appropriate way to describe what students do and do not know about the information they are learning. This concern has both practical and theoretical importance, and it involves issues that overlap three different areas of research: curriculum, educational psychology, and instruction. Because of this diversity of interest, rather different approaches to the problem might be expected, and overall such is clearly the case.

The authors of these chapters report on research being conducted in many different countries. For the most part, they represent the same general approach, although the goals they seek are rather different. Gilbert, Watts, and Osborne (Chapter 2) and West, Fensham, and Garrard (Chapter 3) report on specific techniques for describing cognitive structure. White (Chapter 4) and Pines (Chapter 7) are concerned with the need to organize cognitive-structure data in some meaningful way; White approaches this concern primarily from a psychological and empirical perspective, while Pines is more philosophical and theoretical in nature. Head and Sutton (Chapter 6) remind us of the importance of including affective factors in our concern for representing cognitive structure, and Champagne, Klopfer, and Gunstone (Chapter 5) focus on the extremely important problem of how instruction relates to changes in cognitive structure. Several of the chapters accomplish other purposes in addition to the primary one listed above, and overall the chapters compliment one another in very useful ways.

The primary purpose of this chapter is to provide a critique of the six preceding chapters. In undertaking this task, however, an attempt will be made to accomplish several additional goals in order to provide an appropriate context for understanding the individual chapters. These goals include: (1) providing a general overview of the issues addressed in the various chapters, (2) identifying the commonalities among the various chapters, (3) integrating the various findings in some meaningful way, and (4) relating these findings to school learning.

There are, of course, many ways to approach the problem that these authors have undertaken, and the topic has a long history in both psychology and education (e.g., psychometrics, the study of human memory, etc.). While the present chapters share many of these historical concerns, their approach is very different. The most noticable difference is that all of them reflect the cognitive revolution that has occurred in psychology and education since the early 1970s; consequently, they stand in rather sharp contrast to the more traditional psychometric techniques that are available for describing what students know.

Before something is measured, especially something as complex as cognitive structure, one should have a pretty good understanding of what it is that they are trying to measure. Otherwise, the methodology could become the proverbial tail that wags the dog. Consequently, issues such as the psychological nature of knowledge, the purposes which one might have for measuring cognitive structure, and some general methodological concerns will be discussed. It is hoped that this discussion will aid in developing reasonable criteria against which the chapters (both individually and as a whole) can be evaluated.

THE PSYCHOLOGICAL NATURE OF KNOWLEDGE

Both Pines and West et al. make the point that a combination of epistemological and psychological issues must be dealt with when studying cognitive structure. For years, psychologists and philosophers have pursued a variety of issues concerned with the locus and the nature of knowledge. As used here, *locus* refers to the debate as to whether knowledge exists as an objective entity in the real world or as a psychological entity in the minds of individuals (either collectively or individually). The nature of knowledge refers to the different types of knowledge that exist, regardless of where the knowledge is located. Both of these issues are relevant to our present concerns and will be discussed in turn.

Many individuals tend to think of knowledge as an objective collection of facts and relationships which together comprise a particular subject matter discipline. This "discipline" is thought to exist in books and encyclopedias and to constitute the knowledge that is taught to students in school. Others, most notably psychologists, view knowledge as something that exists either primarily or exclusively in the minds of individual people—a discipline merely represents a group of individuals (i.e., experts in that discipline) who have extremely similar mental representations. The distinction is basically one between the logical and conceptual aspects of a subject matter discipline, on the one hand, and the psychological and cognitive aspects of knowledge representation, on the other. That this issue is still current and unresolved is evidenced by the recent debate in the *Educational Psychologist* between Phillips (1983), Greeno (1983), and Shavelson (1983).

As it turns out, however, there is no *one* body of knowledge. Knowledge about a particular topic can reside in several different locations, and while these bodies of knowledge are related, each one is different from the others in important ways. When we try to measure cognitive structure, we should know which one we are dealing with.

Gilbert et al. identify five "locations" where different bodies of knowledge (science in this case) exist: (1) scientists' science, (2) curricular science, (3) teachers' science, (4) children's science, and (5) students' science. Their taxonomy is the most complete delineation I have seen, and it appropriately represents the various sources of knowledge that must be considered when seeking an adequate understanding of knowledge representation in the context of school learning. Similar, although less comprehensive, distinctions are made between public knowledge and private understandings by West et al. and among the knowledge representations of uninstructed students, novices, and experts by Champagne et al. Both qualitative and quantitative differences exist among these various sources of knowledge, and these differences must be related to one another in meaningful ways if one is to do the most effective job of teaching possible.

The second concern that needs to be considered in attempting to describe cognitive structure has to do with the different types of knowledge that can exist within each of the various loci discussed above. Although at times there is a tendency to think of all knowledge as being basically the same, this way of thinking is beginning to change. In fact, a variety of psychologists, educators, and philosophers have been suggesting for some years that there are several, basically different types of knowledge. Within psychology at the present time, the most commonly encountered distinction is between procedural knowledge (sometimes referred to as algorithmic knowledge) and propositional knowledge (also referred to as declarative knowledge or semantic knowledge). Several of the present authors (White; West et al.; Champagne et al.) make this distinction in some way, employing one of the terms listed above.

The modern-day distinction between propositional and procedural knowledge is often traced to the distinction made by Gilbert Ryle (1949) between "knowing what" and "knowing how." Generally speaking, procedural knowledge refers to one's ability to perform the various procedures necessary to perform some task. The task can be either psychomotor or intellectual. Gagné (1977; Gagné & White, 1978) refers to this type of knowledge as an *intellectual skill,* a term that emphasizes that procedures can be mental as well as motor, although Gagné (1977) also identifies motor skill as a separate type of learning outcome. Propositional knowledge, on the other hand, refers to the systematic and organized body of knowledge that we have about something, and nearly all attempts to describe cognitive structure have focused on propositional (semantic) knowledge. Knowledge, however, clearly is not limited to these two types. For example, Gagné and White (1978) identify four types of organized memory structures: (1) intellectual skills, (2) networks of propositions (propositional knowledge), (3) images, and (4) episodes (autobiographical and temporally related information). These distinctions are discussed in several of the preceding chapters, and White uses them as the basis for identifying various dimensions of cognitive structure.

Within education, there also have been several attempts to distinguish among various types of knowledge, either implicitly or explicitly. Probably the two most commonly encountered systems are the so-called Bloom taxonomies of educational objectives (e.g., Bloom, Engelhart, Furst, Hill, & Krathwohl, 1956) and the various types of learning outcomes identified by Robert Gagné (e.g., 1962, 1977; Gagné & White, 1978). Piaget's (1972) theory of cognitive development also could be cited as an example that bridges both psychology and education.

Being aware that there are different types of knowledge has important implications for both our theoretical understanding of how knowledge is represented and for educational practices. Acquisition of one type of knowledge, for example, does not automatically enable that person to perform a related task involving a different type of knowledge; for example, learning about something (i.e., propositional knowledge) does not mean that the learner will also be able to apply that information in a procedural manner, since a fundamentally different type of knowledge is involved.

Although there have been various approaches to the measurement of knowledge, especially cognitive structure, concern for the most appropriate ways to characterize knowledge for instructional purposes has received very little, if any, attention. Some of these issues will be dealt with in later sections of this chapter, but before these issues are discussed, some consideration must be given to the way in which the appropriateness of any set of measurement procedures depends upon the purpose one has for wanting to do the measurement in the first place.

PURPOSES FOR DESCRIBING STUDENTS' KNOWLEDGE

The purpose one has for wanting to measure something provides the most appropriate and valid source of criteria for evaluating the particular approach being followed. Different approaches often are appropriate for some situations and inappropriate for other situations, depending on the goal one is trying to accomplish. Basically, there are two general purposes for wanting to describe cognitive structure—one research oriented, the other instructionally oriented.

Many individuals interested in measuring cognitive structure are concerned primarily with obtaining a better understanding of how the human mind operates, how students learn, what knowledge students possess, how knowledge evolves psychologically, and many other similar types of concerns. These more or less traditional research concerns generally are descriptive rather than prescriptive in nature (Shuell, 1982) and the practical consequences of the research tends to be long-term rather than short-term. This type of scholarly and scientific understanding has a perfectly legitimate role within education, but the nature of this endeavor needs to be understood if we are to profit most from what it has to offer. Most of the present chapters reflect this purpose as their primary concern, and this focus is especially evident in the chapters by Pines, White, West et al., and Head and Sutton.

Instructional purposes include concerns about performance assessment and/or the making of instructional prescriptions. The former involves information about what students know and the extent to which they are making progress toward some instructional goal or objective. The latter concern involves issues that have to do with the manner in which teachers can or should use information about the knowledge their students have in order to make instructional decisions. The concern here is more for prescriptive rather than descriptive knowledge (Shuell, 1982) about cognitive structure. Although the distinction is not completely clear cut, the Gilbert et al. chapter and especially the Champagne, Klopfer, and Gunstone chapter seem to focus more on this instructional purpose than do the other chapters.

THE MEASUREMENT OF COGNITIVE STRUCTURE

For many years, attempts to describe (i.e., measure) students' cognitive knowledge have employed standard psychometric procedures. This approach is based on several assumptions including: (1) a model of knowledge based on associations and stimulus-response relationships, (2) a focus on the quantitative rather than the qualitative aspects of knowledge (i.e., measuring amount learned rather than extent), and (3) a methodology that concentrates on statistical factors (e.g., normal distributions, etc.) and a particular format (i.e., multiple-choice questions). It would also be fair to say that while the issue was never dealt with in an explicit manner, the assumption generally was made that knowledge exists as subject-matter disciplines rather than psychological entities, although there certainly has been a tremendous sensitivity to the psychological aspects of measuring knowledge. The psychometric approach has proven to be very effective for many situations, and the procedures that have been developed clearly represent one of the major advancements made in the fields of psychology and education during the past century.

Gradually, however, we began to realize the limitations of these assumptions and the inability of traditional psychometric procedures to capture the complexity of certain types of knowledge (i.e., cognitive structure). As a result, newer, more appropriate procedures for representing knowledge and describing the cognitive structure of students were developed. Nevertheless, many of the methodological concerns of the psychometric approach are sound and need to be considered in evaluating these newer approaches. In addition, any comprehensive model of cognitive structure must be able to explain associations and associative data (since they clearly exist) as well as the more complex relationships characteristic of cognitive structure.

Several different approaches to the measurement of cognitive structure can be identified. Speller (1983) has identified five, basically different models of knowledge representation that differ in terms of their "(a) overall organizational patterns, (b) the kinds of information represented, (c) the units of representation (format), and (d) the nature of the relationships among the units of information" (p. 13). These five models are (1) concept models, (2) propositional grammar models, (3) schema models, (4) algorithm models, and (5) computer simulation.

Although a variety of specific techniques for measuring cognitive structure are available, no further attempt will be made here to summarize these methods. Summaries of techniques currently in vogue, including those described in these chapters, are available elsewhere (e.g., Driver & Erickson, 1983; Sutton, 1980; Sutton & West, 1982). Concern for integrating these approaches and summarizing the data they produce is evident in the preceding chapters with, for example, White's concern for identifying dimensions of cognitive structure and Pines' concerns for developing a taxonomy of conceptual relationships. It should be noted that the preceding chapters tend to focus on content (i.e., propositional knowledge) rather than process (i.e., procedural knowledge) and to be concerned with interview techniques for measuring cognitive structure.

METHODOLOGICAL ISSUES

Two general concerns are worth discussing here with regard to methodology: (1) limitations of the data collecting techniques, and (2) ways in which the techniques might be made teacher useable (analogous to teacher-made tests). It might be noted that these two concerns are at least roughly parallel to the two purposes for measuring cognitive structure identified earlier.

While interview techniques can be very effective tools, one must (as is the case with any measurement procedure) exhibit some concern about their replicability. This comment is not meant to suggest a lack of care or awareness on the part of the authors of the preceding chapters but to identify a legitimate limitation inherent in any approach. With care this limitation can be taken into account in such a manner that the usefulness of the technique can be preserved. The difficulties with this technique include, for example, the fact that students often are not able to articulate what they know and what they do not know. Nevertheless, as Ericsson and Simon (1980) point out, verbal reports are data which must be accounted for, and the psychological mechanisms that produce these data must be explained in the same manner that we try to explain other forms of behavior, including performance on psychometric tests. Individuals who use these techniques— or any methodology, for that matter—should be aware of the potential problems, as well as the potential advantages, that are characteristic of that methodology. It is only with this awareness that meaningful conclusions can be drawn.

While interviews are legitimate means of collecting data and may be the most appropriate method for achieving certain purposes, other methods should not be overlooked or categorically rejected. Each technique has certain advantages and disadvantages, and use of any particular one involves trade-offs between these advantages and disadvantages. A combination of methods often provides us with our best understanding of a phenomenon. For example, multiple-choice tests and other "objective" procedures have certain advantages in addition to the limitations discussed earlier, and we sometimes forget that with careful development they are capable of measuring certain aspects of more complex cognitive behavior not totally unlike the relationships encountered in measuring cognitive structure.

For example, Anderson, Reynolds, Schallert, and Goetz (1977) used carefully designed multiple-choice questions to distinguish which of two alternate schemata individuals were using to comprehend a prose passage—the responses to the various questions were selected so that one pattern of responses would indicate the use of one schema while a different pattern would indicate use of the other schema. Likewise, Siegler (1980) used children's response patterns to complex cognitive tasks (including errors) to determine their level of understanding of various concepts and rules. The task was analyzed and the permissible responses that the children could make were constructed in such a manner that one could determine the cognitive rules that they were using to respond to the task (i.e., one rule would produce a particular series of responses while a different rule would produce a different pattern). Thus, more objective techniques can be used to assess at least some of the complexities of cognitive structure, and procedures (such as the I.A.I. technique discussed by Gilbert et al.) that try to combine the advantages of both the more objective and the more subjective approaches have also been developed.

The difficulty involved in the use of more subjective techniques lies not with the inappropriateness of the traditional concerns of psychometrics (e.g., objectivity, reliability, validity, and so forth), but in how these concerns might be meaningfully realized for the situation in which one is interested—in this case, measuring cognitive structure. Some of these concerns for meaningful scientific measurement have been addressed, at least to some extent, in the present chapters. For example, Gilbert, Watts, and Osborne discuss issues associated with reliability and validity, while White identifies three "tests" that should be passed by any set of dimensions used for characterizing cognitive structure, namely, (1) practicality, (2) robustness, and (3) creativity. These types of concerns need to be addressed more systematically if viable techniques for measuring cognitive structure are to be realized.

One issue concerning the measurement of cognitive structure has to do with the extent to which the procedures can be used by teachers. This concern need not be a general requirement (it is perfectly appropriate, as we have already seen, to have purposes that are more research than instructionally oriented), but if any kind of practical utility is desired, at some point, consideration must be given to teachers' ability to use appropriate assessment procedures. Teacher-made tests, as well as standardized tests, were considered to be a legitimate concern of traditional psychometrics, and the same should be true of the overall attempt to describe the cognitive structure of students. Appropriate methods to measure cognitive structure can provide valuable feedback to teachers and can assist in the teaching process itself. While this goal may not be fully realized for some time to come, it is not too early to begin thinking about it. It is encouraging to note that efforts to inform teachers about the relevance of these techniques has already begun (see, for example, Gilbert et al.).

Data collected from the open-ended type of procedures discussed in preceding chapters can be very diverse and difficult to interpret. In order to make sense out of this type of data, some systematic and meaningful system for organizing the data is needed. Scientific investigation depends on this type of data reduction, although *any* system of data reduction emphasizes certain aspects of the raw data while ignoring other aspects. While this inevitable limitation suggests some caution in advocating one particular approach, we nevertheless must seek methods that best suit the purposes we are trying to achieve.

Four of the chapters, with somewhat different purposes in mind, explicitly suggest categories or dimensions for summarizing cognitive structure data. White suggests nine dimensions based on the Gagné and White (1978) model of memory structures. These dimensions are concerned with such things as the extent, precision, and variety of the students' knowledge structure. West et al. suggest five different dimensions concerned with such things as the integration, differentiation, and depth of knowledge. The West et al. dimensions arose out of their attempt to find a meaningful way of describing cognitive structure—an approach that is primarily "bottom-up" when compared with the more theoretical approach taken by White. Nevertheless, the two approaches have a number of similarities, and although data are presented in support of both sets of dimensions, it is extremely difficult, if not impossible, to determine at the present time which if either approach is the most appropriate way of characterizing cognitive structure.

A somewhat different approach is taken by Gilbert et al. These investigators identify "five distinctive types of understanding" concerned with the "types of talk and function" evident in interviews about the meaning of various scientific words. These five types—personal, task, card, concept, and framework—appear to be primarily functional in nature, although the authors do not provide a rationale for their selection. Finally, Pines suggests two fundamental types of conceptual relationships: set-subset relationships and whole-part relationships. The ultimate usefulness of these various systems for characterizing cognitive structure will depend on their ability to relate to and help explain other types of psychological and instructional variables.

KNOWLEDGE REPRESENTATION AND SCHOOL LEARNING

As emphasized in several of the chapters in Part I, school learning involves more than the mere acquisition of many isolated facts—although we all know of classrooms where this appears to be the main objective. Rather, most people view education as the means by which students develop and/or change complex cognitive structures involving not only facts but also relationships which tie the facts together into meaningful wholes. The important educational question, then, becomes one of how we can help students acquire appropriate cognitive structures or change their existing ones in appropriate ways. This task is not nearly as simple as it may seem. As Shavelson (1981) points out, for example, in teaching any subject matter there is a need to address four related topics: (1) the subject matter being taught, (2) the student, (3) the teacher, and (4) the instructional context. The bottom line, however, is student learning, and consideration must be given to how this can occur in the most productive manner.

In recent years, there has been a change in the way people think about learning in general and school learning in particular. In many ways this change has accompanied the so-called cognitive revolution in psychology. At the present time, learning is generally considered to be an active, constructive process rather than a passive, reproductive process (as was the general case in the 1960s and early 1970s). Although some individuals, of course, have long considered learning to be an active process, this orientation did not become part of the mainstream of psychological thinking until fairly recently.

While this change in conceptualization was taking place, however, those individuals most interested in cognitive structure, knowledge representation, and cognitive psychology showed little concern for learning as such— that is, concern for the factors or variables that influence changes in human performance, knowledge structures and/or cognitive conceptions. This situation is gradually beginning to change. Probably the first modern attempt to develop a cognitive conception of learning was Rumelhart and Norman's (1978) logical analysis of the nature of learning within a schema-based representational system. According to Rumelhart and Norman, there are three qualitatively different kinds of learning:

1. Accretion, or the encoding of new information in terms of existing schemata;
2. Tuning or schema evolution, which involves the slow modification and refinement of a schema as a result of using it in different situations; and
3. Restructuring or schema creation, which is the process whereby new schemata are created.

Meaningful school learning involves all three types, with different types of learning being most appropriate in different instructional situations. Each type must be taken into account in developing an adequate understanding of cognitive structure and its relationship to school learning.

As Gilbert et al. point out and Champagne, Klopfer, and Gunstone clearly document, students enter a classroom with preexisting conceptions that determine how the instructional material and events encountered in the classroom are interpreted. This fact is not too surprising when one realizes that all individuals are active learners trying to make sense out of the world and events that surround them. As a result of their many experiences, students have developed their own conceptions before they ever enter a classroom, and in many cases the individual has found these conceptions to be quite successful. It is important to remember, especially with regard to school learning, that all learning is cumulative—no learning occurs in isolation.

At times, the students' preexisting conceptions are consistent with what is being taught, but at other times—as Champagne, Klopfer, and Gunstone demonstrate—these preexisting conceptions can be antagonistic to the conception that the teacher is trying to present, and these preexisting conceptions can be extremely difficult to change. The importance of this point is clearly evident in the following statement by Champagne, Klopfer, and Gunstone:

> Students misunderstand physics instruction because they interpret their physics lectures and texts in the context of their real-world definitions of the terms, rather than in the way that the text or teacher is using the terms. For example, students use the terms, speed, velocity, and acceleration interchangeably.

This realization has very important implications for instruction.

What then, can be done to help students change their cognitive structures in appropriate ways? Anderson (1977) discusses several ways in which it might be possible to make substantial changes in higher level conceptual frameworks of the type being discussed here, including Socratic questioning and forcing students to confront difficulties in their current way of thinking and helping them to discover the ability of a different schema to resolve these difficulties. The more specific instructional strategies suggested by Champagne, Klopfer, and Gunstone represent an important step forward since the procedures are derived from a logical and theoretical analysis of several factors involved in the acquisition or modification of knowledge structures. The viability of these procedures for meaningful learning in the classroom is clearly a matter worth pursuing further.

One final concern that needs to be addressed in discussing the relationship between knowledge representation and school learning is found in the distinction among the various sources of knowledge made by Gilbert et al. These various sources must be taken into account when considering classroom learning, and the ability to coordinate them has important implications for both curriculum planning and instructional planning. The assessment of what students know relative to what we want them to know provides information that can be used for both student evaluation and/or diagnosis (i.e., what knowledge does the student need to acquire next), but this process can be effective only if we keep in mind the various types of knowledge that we are dealing with and the various instructional procedures that can be used to help students learn the desired knowledge.

IMPLICATIONS FOR FUTURE RESEARCH

The preceding chapters have a number of important implications for future research. Some of these implications reflect limitations of current research on cognitive structure, while other implications build upon and extend current findings. It must be recognized that any approach to the measurement of knowledge makes certain assumptions about the nature of knowledge. This is true whether, for example, traditional psychometric procedures are used to measure knowledge which is thought to consist of associative bonds, or cognitive mapping procedures are used to represent complex relationships among various aspects of cognitive structure. These assumptions need to be challenged in an attempt to determine the most appropriate way (or ways) of representing knowledge for instructional purposes. Do some measurement approaches, for example, relate better than others (either empirically or theoretically) to various factors related to instruction (e.g., student characteristics, different types of instructional outcomes, instructional/learning variables, and so forth)?

Another concern worth pursuing has to do with the advantages and disadvantages of thinking about the qualitative and the quantitative aspects of knowledge. Related to this concern are issues involving the soundness of scientific data. The traditional concepts of reliability and validity, for example, may be difficult to implement in some situations, but the concepts are sound ones nevertheless, and methods to deal with the concerns they represent must be found if our investigations are to remain based on sound data. Science transcends any given paradigm or method, and as our investigations move into new areas, we must be creative in meeting the challenge of finding new ways to validate our results.

Another general concern for future research involves the role of *instructional* variables. How exactly, for example, does one help a novice to become an expert? Champagne, Klopfer, and Gunstone make a beginning in this direction, and the effort should continue. Along this same line, several of the chapters suggest that learning should be meaningful, but what does a teacher do to help make classroom learning a meaningful experience? We certainly know something about this task, but much more information is needed before we have a clear understanding of exactly how meaningful learning can be facilitated. For example, is it sufficient to present missing knowledge to a student once we have discovered what he or she already knows, or should we try to elicit certain learning processes, perhaps by employing an advance organizer? (Ausubel, 1968). We have little precise knowledge in this regard that is capable of guiding a teacher's decision about what to do next (Shuell, 1981).

Finally, it was noted earlier in this chapter that the acquisition of one type of knowledge does not automatically provide one with the ability to perform a related task that involves a different type of knowledge; for example, learning about something (propositional knowledge) does not mean that the learner will also be able to use that information in a procedural manner since application involves a fundamentally different type of knowledge. This realization has important implications for research concerned with the retention and transfer of knowledge. An individual's cognitive structure involves both (and perhaps other) types of knowledge, and we need to investigate various ways of representing knowledge in order to determine the extent to which they can be appropriately related to methods of instruction known to be effective in helping students to acquire, retain, and transfer the large amount of knowledge presented in school.

CONCLUSION

This chapter has addressed a number of general concerns associated with the measurement of cognitive structure and its relationship to school learning. Its main purpose has been to discuss the six preceding chapters in this ·section of the book and to place their contributions into an appropriate context. Together the chapters represent a useful state of the art report on the measurement of cognitive structure and school-type knowledge. Hopefully, they will provide a stimulus to further advances in our theoretical understanding of school learning and how that learning can be facilitated through effective teaching.

REFERENCES

Anderson, J. R. (1982). Acquisition of cognitive skill. *Psychological Review, 89,* 369–406.
Anderson, R. C. (1977). The notion of schemata and the educational enterprise: General discussion of the conference. In R. C. Anderson, R. J. Spiro, & W. E. Montague (Eds.), *Schooling and the acquisition of knowledge.* Hillsdale, NJ: Erlbaum.
Anderson, R. C., Reynolds, R. E., Schallert, D. L., & Goetz, E. T. (1977). Frameworks for comprehending discourse. *American Educational Research Journal, 14,* 367–381.

Ausubel, D. P. (1968). *Education psychology: A cognitive view.* New York: Holt, Rinehart, & Winston.

Bloom, B. S., Engelhart, M. D., Furst, E. J., Hill, W. H., & Krathwohl, D. R. (1956). *Taxonomy of educational objectives. Handbook I: Cognitive domain.* New York: McKay.

Driver, R., & Erickson, G. (1983). Theories-in-action: Some theoretical and empirical issues in the study of students; conceptual frameworks in science. *Studies in Science Education, 10.*

Ericsson, K. A. & Simon, H. A. (1980). Verbal reports as data. *Psychological Review, 87,* 215–251.

Gagné, R. M. (1962). The acquisition of knowledge. *Psychological Review, 69,* 355–365.

Gagné, R. M. (1977). *The conditions of learning* (3rd ed.). New York: Holt, Rinehart, & Winston.

Gagné, R. M. & White, R. T. (1978). Memory structures and learning outcomes. *Review of Educational Research, 48,* 187–222.

Greeno, J. G. (1983). Response to Phillips. *Educational Psychologist, 18,* 75–80.

Piaget, J. (1972). *The principles of genetic epistemology* (W. Mays, Trans.) New York: Basic Books.

Phillips, D. C. (1983). On describing a student's cognitive structure. *Educational Psychologist, 18,* 59–74.

Rumelhart, D. E. & Norman, D. A. (1978). Accretion, tuning, and restructuring: Three modes of learning. In J. W. Cotton & R. Klatzky (Eds.), *Semantic factors in cognition.* Hillsdale, NJ: Erlbaum.

Ryle, G. (1949). *The concept of mind.* London: Hutchinson.

Shavelson, R. J. (1981). Teaching mathematics: Contributions of cognitive research. *Educational Psychologist, 16,* 23–44.

Shavelson, R. J. (1983). On quagmires, philosophical and otherwise: A reply to Phillips. *Educational Psychologist, 18,* 81–87.

Shuell, T. J. (1981). Toward a model of learning from instruction. In K. Block, *Psychological theory and educational practices: Is it possible to bridge the gap?* Symposium presented at the meeting of the American Educational Research Association, Los Angeles, CA.

Shuell, T. J. (1982). Developing a viable link between scientific psychology and educational practices. *Instructional Science, 11,* 155–167.

Siegler, R. S. (1980). When do children learn? The relationship between existing knowledge and learning. *Educational Psychologist, 15,* 135–150.

Speller, K. R. (1983). *Learning ability and learning problems: Four proposed learning traits and an approach to their investigation.* Unpublished doctoral dissertation, State University of New York at Buffalo.

Sutton, C. R. (1980). The learner's prior knowledge: A critical review of techniques for probing its organization. *European Journal of Science Education, 2,* 107–120.

Sutton, C. R. & West, L. H. T. (1982). *Investigating childrens' existing ideas about science.* Occasional Publication, University of Leicester, School of Education.

STABILITY
AND
CHANGE
IN CONCEPTUAL
UNDERSTANDING

ACQUIRING AN EFFECTIVE
UNDERSTANDING OF SCIENTIFIC
CONCEPTS*

F. Reif

In quantitative sciences, such as physics, special concepts and associated principles are logically the basic building blocks of the knowledge used to deduce important consequences, make predictions, and solve problems. However, mere definitions of concepts or statements of principles are psychologically far too primitive building blocks to permit the performance of complex intellectual tasks.

To be functionally useful, a conceptual building block (or "concept schema") must include a concept accompanied by the ancillary knowledge needed to make the concept effectively usable. In particular, this knowledge must be sufficient to ensure that the concept can be used *reliably*, that is, without errors or ambiguities; *easily* and *rapidly*, so that use of the concept leaves adequate attention and time available to deal with other aspects of complex tasks; and *flexibly*, so that the concept can be used reliably in diverse and unfamiliar contexts. Similar comments can be made about a principle relating previously defined concepts.

The ancillary knowledge, required to make a concept or principle effectively usable, is far from trivial. Striking evidence supporting this statement comes from several recent studies (e.g., Viennot, 1979; Trowbridge and McDermott, 1981; Clement, 1982; di Sessa, 1982). These show that many students, after having studied physics concepts and having been familiar with them for an appreciable time, may nevertheless lack the ancillary knowledge needed to use such concepts reliably. Correspondingly, they exhibit major misconceptions and errors.

The preceding comments indicate the importance of analyzing and explicating the ancillary knowledge required to make a scientific concept or

*This work was partially supported by the National Science Foundation under Grant No. SED 79-20592. I am indebted to Joan I. Heller for useful comments.

principle affectively usable. Such an analysis, discussed in this paper, is interesting and useful from several points of view:

1.   From a scientific or psychological point of view, such an analysis helps make explicit underlying knowledge which is necessary (although *not* sufficient) for any scientific problem solving. (Reif 1981; Reif & Heller, 1982). It also helps reveal important knowledge which is often "tacit", i.e., which is possessed by experts without their conscious awareness of its existence. Finally, such an analysis can help to predict many of the difficulties and errors exhibited by inexperienced students.

2.   From the practical perspective of teachers, such an analysis can help to identify important knowledge essential to students' understanding and learning of concepts or principles. Accordingly, it can be useful for diagnosing and minimizing the difficulties experienced by many students. Furthermore, it can provide the basis of explicit instructional methods for teaching concepts or principles more effectively.

3.   From the practical perspective of students, such an analysis can provide guidelines for studying concepts more effectively and can thus help students to acquire some important general learning skills.

As the analysis in the following pages indicates, the basic ancillary knowledge required to make a concept or principle effectively usable is remarkably large (although it is commonly possessed by any expert). This is one reason why the learning of a new scientific concept is a difficult task for students.

KINDS OF CONCEPTS AND ASSOCIATED ANCILLARY KNOWLEDGE

The simplest kind of concept is a particular "entity" (e.g., "the sun"). Any member of a specified set of entities (e.g., "triangle," "particle") is then a "generic concept" or "variable."

A "property" is a more complex kind of concept used to describe one or more other concepts called the "independent variables" described by the property. This description is achieved by associating a unique value of the property for any possible set of values of the independent variables.[1] (If the property is a "quantity", its values are numbers; otherwise they may be members of any other specified set.) For example, "area" is a property describing a surface by associating a particular value (a positive real number) to any member of a set of entities called "surfaces." Similarly, "color" is a

[1]A property is thus, in a generalized sense, a mathematical function associating a unique value of a variable to any set of values of some independent variables.

TABLE 9.1   Interpretation of a Concept

Specification
Specification of concept
    Summary description
    Informal description
    Procedural specification
    Applicability conditions
Specification of concept values
    Ingredients and symbolic expression (elements specifying type, units)
    Possible values (and typical values)
Specification of independent variables
    Basic independent variables and symbolic expression
    Relevant properties of independent variables

Instantiation
    Various values of independent variables and of their properties
    Various symbolic representations

Error prevention
    Warnings about likely errors (see Table 2)
    Discrimination between each error and correct case
    Helpful symbolism

---

property describing an object by associating a particular value (one of the set of concepts "red," "yellow," "green," . . .) to any object. As a last example, "velocity" is a concept describing jointly a particle, a reference frame, and a time by associating a vectorial numerical value to any particle for any reference frame and for any time. The particular independent variables described by a property are indicated by appropriate prepositions: for example, one speaks of the area *of* a surface; or of the color *of* an object; or of the velocity *of* a particle relative *to* a particular reference frame *at* some specified time.

The discussion in the following pages will deal predominantly with property concepts, because these are centrally important to provide the descriptions needed in any science. The analysis of the ancillary knowledge required to make a property concept effectively usable includes, as a subset, the ancillary knowledge for a simple entity concept. Furthermore, as discussed later, the ancillary knowledge for a property concept is essentially the same as that for a principle.

The most important ancillary knowledge, required to make a concept effectively usable, is that required to *interpret* the concept appropriately. This knowledge, summarized in Table 9.1 and discussed in the next three sections, includes that needed to specify the concept, to achieve this specification in various particular instances, and to do this without committing errors of interpretation. The other kinds of ancillary knowledge (e.g., knowledge about basic implications, knowledge about alternative symbolic representations, and guidelines about when and how to use the concept) will not be discussed further in this paper. However, the analysis of the ancillary knowledge needed for concept interpretation will be used to point out some practical implications for the teaching of scientific concepts or principles.

SPECIFICATION KNOWLEDGE

As indicated in Table 9.1, the most basic knowledge required to interpret a scientific concept is that needed to specify the concept fully and unambiguously. The important components of this "specification knowledge" are now discussed in turn.

SPECIFICATION OF A CONCEPT

Ultimately, the meaning of any scientific concept must be specified by explicit rules (e.g., definitions) which ensure that the concept is unambiguously identified so that it can lead to clearly interpretable scientific knowledge. The following ways of specifying a concept are *all* useful—summary descriptions because they are compact and easily remembered, informal descriptions because they clarify the essential meaning of a concept, and procedural specifications because they provide the most detailed specification.

Summary Description

A summary description of a concept is useful because it provides a brief and precise statement of the meaning of the concept, a statement which can be easily remembered and used as the starting point for more complete elaborations. A typical example of such a summary description is the formal statement $a = dv/dt$ which defines compactly the concept "acceleration" (denoted by a) in terms of the velocity v and the time t.

Informal Description

An informal description of a concept is useful because it specifies the essential meaning of a concept without undue precision or excessive details. By focusing attention selectively on a few salient features, an informal description can help in relating a concept to more familiar knowledge and in retrieving the concept in complex situations. Indeed, such qualitative informal descriptions (and methods of successive refinement which proceed from qualitative to more detailed descriptions) can be very useful in facilitating problem-solving tasks. (Larkin and Reif, 1979).

For example, the acceleration of a particle may be described informally by statements such as "acceleration is the rate of change of velocity with time" or "acceleration is a quantity describing the small change of a particle's velocity during a small time." Such statements are admittedly rather vague, but they make quite clear what essential quantities are interrelated by the property "acceleration" and when this property might be relevant.

Procedural Specification

The preceding specifications of a concept, whether formal or informal, are descriptive (or "declarative"); that is, they are expressed in terms of state-ments asserted to be true. A very important alternative way of specifying a concept is by means of a step-by-step *procedure* specifying how to identify or exhibit the concept Such a "procedural specification" provides the most explicit and detailed specification of a concept. It also has fundamental scien-tific importance as an operational definition which specifies what one must actually *do* to decide whether a concept is properly identified.

These remarks can be exemplified by the following procedural specifica-tion of the concept "acceleration": (1) Consider a specified particle P. (2) At some specified time t, consider the velocity of v of P relative to some spec-ified reference frame R. (3) For comparison, consider some neighboring time $t' = t + \Delta t$ and consider the velocity v' of the particle P at this time. (4) Find the velocity change $\Delta v = v' - v$ by subtracting vectorially the old velocity v from the new velocity v'. (See Fig. 9.1.) (5) Calculate the ratio $\Delta v/\Delta t$. (6) Verify that the time t' has been chosen sufficiently close to t so that a closer choice, making Δt smaller, would leave the ratio $\Delta v/\Delta t$ unchanged within the desired precision of description. In this case denote Δt by dt and Δv by dv. (7) Identify the resulting ratio as the concept of interest and name it the "acceleration of P relative to R at the time t."

The preceding procedural specification makes abundantly clear the many complexities involved in the definition of the concept "acceleration," com-plexities which are largely hidden in the formal descriptive specification $a = dv/dt$. Indeed, the distinction between a procedural specification and a formal description is strikingly apparent in practice. For example, when students are asked to find the acceleration of a pendulum bob at the extreme position of its swing where its velocity is zero, many students say that the

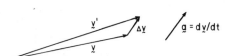

FIGURE 9.1   Velocity change and acceleration.

acceleration is zero. Most of them continue to make this claim vociferously, even when they are specifically asked to use the definition of acceleration, written out explicitly as $a = dv/dt = (v' - v)/(t' - t)$. But when these students are asked to follow the steps of the *procedure* specifying the acceleration, they change their minds and realize that the acceleration is nonzero. (Of course, experts are much more skilled in translating a formal description into a corresponding procedure.)

As another example, when novice students are asked to find the component of a vector **V** along some specified direction **i**, most can easily answer this question when the direction **i** is horizontal, as shown in Figure 9.2a. On the other hand, they often have difficulties in more general cases, such as that shown in Figure 9.2b. But such difficulties disappear if students have learned the *procedure* specifying how to identify or find the component of a vector along some given direction. The reason is that such a procedure does *not* merely rely on the recognition of a familiar pattern. (Instead, it identifies the component by the general process of drawing, from the ends of the arrow representing the vector **V**, lines parallel and perpendicular to the given direction **i**.)

As the preceeding examples illustrate, it can be pedagogically very useful if students are asked to explain the meaning of a concept by specifying an appropriate procedure.

Applicability Conditions

A detailed procedural specification helps make apparent the conditions under which a concept may legitimately be applied. Such applicability conditions must be made quite explicit to help avoid misinterpretations and errors.

For example, the concept "acceleration" can be applied to any *particle*, but not indiscriminately to any system of particles (a mistake sometimes committed by students). As another example, the concept "potential energy" must be accompanied by the applicability condition specifying that this

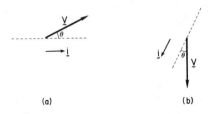

(a)   (b)

FIGURE 9.2   A vector **V** and a direction **i**.

concept can be used only for interactions described by *conservative* forces (i.e., forces which do work independent of the process between states of a system.)

SPECIFICATION OF CONCEPT VALUES

The specification of a concept implies a corresponding specification of its values. Although such a knowledge about values is relatively simple, it needs to be made explicit if errors are to be avoided. Table 9.1 and the following paragraphs outline the most important knowledge about the values of a concept.

Value Ingredients

The value of a concept is ordinarily specified by several ingredients, the *elements* needed to specify the type of value and the *units* needed for specification. For example, the concept "acceleration" has values which are vectors. The elements needed to specify this type of value are a "magnitude" and a "direction." The units are "meter/second^2".

In the case of value specification, as well as in more complex cases discussed later, the use of explicit symbolic expressions is an important aid to ensure correct usage of a concept. For mere adherence to proper symbolic form (or "syntax") helps automatically ensure that a specification is complete and correct. For example, an appropriate symbolic expression for a value of the concept "acceleration" is "⟨magnitude with unit of length/time^2⟩ along ⟨direction⟩." Here anything enclosed between angular brackets indicates a "slot" to be filled by an instance of the specified kind of entity. For example, a *correct* value specification of an acceleration might be "1.6 m/sec^2 along the northern direction". By contrast, a value specification such as "1.6 m/sec^2" would be *incomplete* and thus ambiguous because the slot about direction has not been filled in. Similarly, a value specification such as "1.6 m/sec along the northern direction" would be *incorrect* because the slot for units has been filled by the wrong kind of unit.

Possible Values

Proper value specification requires also knowledge about the domain of possible values of a concept (e.g., knowledge that the concept "kinetic energy" can assume all *non-negative* numerical values.) A knowledge of *typical* values is also valuable for making qualitative predictions and checking the solutions of problems. For example, it is useful to know that typical values of the acceleration have magnitudes of the order of a few meters/second^2 for falling objects or accelerating cars.

SPECIFICATION OF INDEPENDENT VARIABLES

Basic Independent Variables

The specification of a property concept implies a corresponding knowledge of all the basic independent variables needed to specify this property completely. Such knowledge can be subtle and needs to be made explicit to avoid likely errors and ambiguities.

For example, the concept "acceleration" is a property used to describe a particle at some particular time relative to some particular reference frame. Hence a complete specification of the concept "acceleration" requires a specification of *all* the following independent variables, namely "particle," "time", and "reference frame". Failure to specify any of these independent variables leads to ambiguities (i.e., no unique value could then be ascribed to the acceleration, nor could statements about this concept be judged true or false). For instance, the statement that "the acceleration of a ball at some instant is 10 m/sec^2 downward" involves an *incomplete* specification of the acceleration because of failure to specify a reference frame. Thus, the statement is ambiguous; for example, it might be true if the earth is used as a reference frame, but false if the reference frame is an elevator moving relative to the earth.

An explicit knowledge of *all* the basic independent variables needed to specify a concept unambiguously is very important to the proper interpretation of a concept. (Indeed, deficiencies in such knowledge lead to many common confusions observed among students.) The use of explicit symbolic expressions is again a powerful aid for ensuring that a property concept is specified completely and correctly. For example, the word "acceleration," by itself, is really meaningless. Instead, the adequately defined concept is the one denoted by the full expression "the acceleration of ⟨particle⟩ at ⟨time⟩ relative to ⟨reference frame⟩," where each entity between angular brackets denotes a slot to be filled by a variable of the specified kind.

Consistent use of full symbolic or verbal expressions can greatly help students (and occasionally even experts) to avoid fuzzy thinking and thus to prevent many errors or confusions.[2] For example, talking about the "velocity of some ball at some particular time relative to some particular reference frame" focuses explicit attention on all relevant entities. On the other hand, when talking blithely about the "velocity of a ball," students are often lead to assume inappropriately that the velocity if relative to the earth (since

---

[2]A full symbolic expression may, of course, be abbreviated if this is done explicitly *after* the omitted independent variables have been specified once and for all at the beginning of a discussion. For example, one may speak simply of the "velocity of the ball" *after* one has specified that the reference frame is the laboratory.

specification of a reference frame has been ignored) or to assume inappropriately that the velocity is constant (since specification of a particular time has been ignored).

Another example, illustrating the importance of complete specifications, is provided by the concept of "force." In physics, this concept is used to describe the interaction between particles and requires, therefore, the specification of at least *two* particles. Accordingly, the symbolic expression for force is of the form "force on ⟨particle⟩ by ⟨other particle⟩" where it is essential that *both* slots be properly filled. Indeed, to help students avoid errors and confusions, it is very useful to insist that students never use the word "force" unless followed by the phrase "*on . . . by . . . .*" Insistence upon use of this full expression avoids the lay conception of force as an intrinsic property inherent in an object, as expressed by phrases such as "force *of* an object." It helps to avoid confusions between "action" and "reaction" if these historically hallowed words are discarded in favor of the much clearer expressions "force on *A* by *B*" and "force on *B* by *A*." It also helps to avoid students' inappropriate invocation of non-existing "centripetal" or "centrifugal" forces produced by no discernible objects.

Relevant Properties of Independent Variables

As indicated in Table 9.1, it is important to know not only which basic independent variables are needed to specify a given concept, but also which particular properties of these variables are (or are not) required for a complete specification. For example, as mentioned previously, the basic independent variables needed to specify a "force" are the particle *on* which the force acts and the particle *by* which it is exerted. But not all properties of these particles are relevant to this specification. For instance, the positions of the particles are relevant and must be specified. On the other hand, the colors of these particles are irrelevant, as are their velocities (for ordinary central forces).

Note that the preceding knowledge, needed to explicate what particular parameters are (or are not) relevant to a specification of a given concept, is far from trivial. Indeed, it implies important understanding of functional dependencies or invariances in situations where the concept is pertinent.

INSTANTIATION

In principle, the knowledge required to specify a concept adequately, as discussed in the preceding section, is sufficient to interpret the concept. But this knowledge, although essential, is too general and abstract to make the concept effectively usable in practice. Thus, it is also necessary to know how

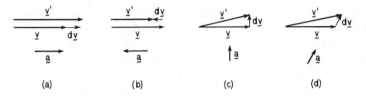

FIGURE 9.3   Various cases of velocity changes.

to "instantiate" the concept, that is, how to apply the concept reliably in various possible kinds of specific instances. (Indeed, it is a familiar fact that many students, even when able to state the definition of a concept, may be quite unable to apply this definition in particular cases.)

As indicated in Table 9.1, the knowledge needed to instantiate a concept involves the ability to do the following: (1) To identify or exhibit the concept for various possible values (or relative values) of the independent variables or of their properties. (2) To do this in various possible symbolic representations, for example, in words, in pictures (diagrams or graphs), or formal mathematical symbolism.

For example, the acceleration $\mathbf{a}$ (defined by $\mathbf{a} = d\mathbf{v}/dt$) involves a comparison of the velocity $\mathbf{v}$ of a particle at some specified time t and of its velocity $\mathbf{v}' = \mathbf{v} + d\mathbf{v}$ at a slightly later time $t' = t + dt$. Adequate instantiation knowledge then requires the ability to apply the concept "acceleration" in the following kinds of cases, described verbally as well as pictorially: (1) The new velocity $\mathbf{v}'$ has the *same direction* as the original velocity $\mathbf{v}$, but a larger or smaller magnitude, as indicated in Figures 9.3a and 9.3b. The acceleration $\mathbf{a}$ has then, respectively, either the same or opposite direction compared to the velocity $\mathbf{v}$. (2) The new velocity $\mathbf{v}'$ has the same magnitude as $\mathbf{v}$, but a different direction, as indicated in Figure 9.3c. The acceleration has then a direction perpendicular to the velocity $\mathbf{v}$. (3) In the most general case, the new velocity $\mathbf{v}'$ differs from $\mathbf{v}$ in both magnitude and direction, as indicated in Figure 9.3d. The acceleration has then a direction *not* parallel to the velocity $\mathbf{v}$, but toward the concave side of the particle's path.

Being able to identify and use various possible instances of a concept is sometimes far from trivial. For example, it often takes students a long time to understand that the innocent-looking definition $\mathbf{a} = d\mathbf{v}/dt$ of the concept "acceleration" encompasses all the various cases illustrated in Figure 9.3.

ERROR PREVENTION

Human beings are prone to errors. The reliable interpretation of a concept requires, therefore, also adequate knowledge to prevent errors, that is, knowledge to avoid likely errors, to detect such errors when they have been committed, and to correct them appropriately.

As indicated in Table 9.1 such error-prevention knowledge includes explicit warnings or "caveats" about errors likely to occur in the application of the concept; knowledge about how to discriminate any such error from the correct situation; and the use of explicit symbolism designed to help avoid such errors.

WARNINGS ABOUT LIKELY ERRORS

Reliable performance on any task is obviously facilitated if one is explicitly forewarned about likely errors and pitfalls. Such errors may be identified by actual observations of commonly made errors. A theoretically more interesting approach is to use an a priori analysis to predict many of the kinds of errors likely to occur in the use of any newly encountered concept. Such an analysis must take into account the characteristics of the particular concept, for example, the previously discussed knowledge required for the specification of the concept. It must also take into account the characteristics of the person using the concept, including the person's preexisting knowledge. The results of such an analysis are briefly outlined in Table 9.2 which indicates some of the most common basic errors likely to occur in the application of any concept.[3]

The likely errors listed in Table 9.2 correspond to errors in the various kinds of specification knowledge summarized in Table 9.1. The following paragraphs discuss and exemplify the most likely of these errors. The first two of these are gross confusions which result if a concept is identified by relying merely on the recognition of some salient features, rather than by applying explicitly the rules specifying the meaning of the concept.

Confusion of a Concept with Another Concept Denoted by a Similar Symbol (Including Lay Terminology)   Such a confusion occurs because a superficial similarity of symbols causes a failure to discriminate between different concepts. For example, the scientific concept "acceleration" (denoting the vector dv/dt describing the vectorial change of velocity) is likely to be confused with the lay term "acceleration" (used in everyday life to denote roughly the rate of increase of speed with time). As another example, concepts such as "kinetic energy," "potential energy," and "energy" may easily be confused because their names all include the same word "energy."

---

[3]Different basic errors listed in Table 9.2 may sometimes lead to the same overtly observable error; that is, the same observable error may sometimes be traceable to different underlying errors.

TABLE 9.2   Likely Errors

Errors in specification of concept
  Gross confusions
    Confusion with concept denoted by similar symbol
    Confusion with concept describing different features of same situation
  Errors in specification rules
  Errors in applicability conditions

Errors in specification of values
  Errors in specifying ingredients
  Errors in possible values

Errors in specification of independent variables
  Omitted independent variables
  Wrong independent variables or properties thereof

Confusion of a Concept With Another Concept Describing a Different Feature of the Same Situation   Such a confusion is caused by a failure to discriminate between related concepts which occur frequently in the same context. For example, "acceleration" and "velocity" are likely to be confused because both these concepts describe the motion of a particle, although different features of such motion.

Errors in Specification Rules   Even if a detailed rule or procedure is used to identify a concept, an error in some *part* of the rule can lead to misidentification of the concept. There may be many such possible errors because one or more steps in a specification rule may be omitted or wrong.

For example, the procedural specification of the concept "acceleration" involves a subtraction $\mathbf{v}' - \mathbf{v}$ of velocities at slightly different times. If this *vectorial* subtraction is confused with a *numerical* subtraction of magnitudes, a wrong concept (the rate of change of speed dv/dt) is identified.

Errors in Applicability Conditions   An example of such an error would be the attempted use of a potential energy to describe interaction due to friction forces (since the concept of potential energy is only applicable in the case of conservative forces).

A particularly common error in applicability conditions occurs when a concept, describing a special case, is inappropriately extended to a more general case where it is not valid. Such confusions of special cases with general cases are particularly likely when the special case has an appealing simplicity and has been encountered first in one's learning experience. For example, students often encounter the concept "velocity" first in the simple special case of uniform motion along a straight line when the velocity may be simply defined by the numerical ratio s/t (where s is the distance traveled during the time t). It is then predictably likely that students will subsequently confuse this definition of the concept with the general concept of "velocity" defined as the *vector* $d\mathbf{r}/dt$ (where $d\mathbf{r}$ is the *infinitesimal* displacement $d\mathbf{r}$ during an *infinitesimal* time dt).

Errors in Specification of Values   Errors in the specification of the values of a concept occur when some of the ingredients necessary to specify a value are omitted or wrong, or because impossible values are attributed to the concept. Such errors are easy to avoid, although common among novice students. The following are examples of such errors: Describing the value of an acceleration by specifying a magnitude without a direction; specifying the value of a potential energy with the wrong unit "newton"; or stating that the value of a kinetic energy is negative.

Errors in Specification of Independent Variables   A very common kind of error results from the omission of some of the independent variables required to specify a property concept. The consequences are an incomplete specification of the concept and concomitant ambiguities; these can often lead to troublesome confusions and seemingly perplexing paradoxes. The following are examples of such omissions: talking about an acceleration without specifying the reference frame relative to which it is measured; talking about a potential energy without specifying the standard position from which it is measured; or talking about a force without specifying the object exerting this force.

DISCRIMINATIONS

Table 9.2 and the preceding comments help to identify likely errors which must be avoided if a concept is to be used reliably. Hence it is essential to be able to discriminate between any such error and the correct application of the concept. To acquire the ability to make such discriminations while learning an unfamiliar concept, it is useful to compare explicitly the error (and its consequences) with the correct situation. Distinguishing features, characterized abstractly as well as exemplified in specific cases, can then be made explicitly apparent so that they can be readily recognized and heeded.

As an example, consider the error involving the confusion of the concept "acceleration" with the concept "velocity." Explicit comparison of these concepts leads to the knowledge needed to discriminate between them. In particular, the two concepts are characterized by the following distinguishing features: the acceleration describes the rate of change of *velocity*, whereas the velocity describes a rate of change of *position;* also the unit of acceleration is *meter/second²*, whereas the unit of velocity is *meter/second.* Specific examples illustrating distinctions between these concepts are the following: the acceleration can be zero while the velocity is non-zero (e.g., for motion with constant velocity); the acceleration can be non-zero while the velocity is zero (e.g., at the highest point of a ball thrown vertically upward); and the acceleration can be constant while the velocity is changing (e.g., for a freely falling object). A knowledge of such discriminations for each likely error is an important part of the ancillary knowledge needed to make a concept reliably usable.

HELPFUL SYMBOLISM

A powerful aid for preventing errors is the introduction and use of appropriate symbolism, for then strict adherence to symbolic form can automatically help to avoid many errors.

As a trivial example, confusion between the concept "velocity" (a vector) and the concept "speed" (the magnitude of the velocity) can be minimized by consistently using the letter v (printed in boldface type or underscored by a squiggly line) to denote the *vector* representing the velocity, while using the unadorned letter v to denote the *number* representing the speed.

Much more important examples of helpful symbolism involve the use of standarized symbolic expressions with "slots" indicating explicitly all the kinds of information that need be supplied. As previously discussed and exemplified, such symbolic expressions can be used to indicate explicitly all the ingredients needed to specify the value of a concept or all the independent variables needed to specify a property. Consistent use of such symbolic forms can greatly help to avoid many errors of omission or commission in the application of concepts.

APPLICATION TO PRINCIPLES

The preceding sections discussed at some length the ancillary knowledge needed to interpret concepts (e.g., properties such as "acceleration," "potential energy," etc.). The preceding discussion can be readily extended to principles expressing important relations between previously defined concepts (e.g., the principle $\Delta K = W$ relating kinetic energy and work, or the gravitational force law $F = Gm_1\,m_2/R^2$).

Indeed, any valid relation between concepts can be regarded as a "truth property" (or "predicate") which asserts that the property has the value "true" whenever the values of the concepts are related in some specified way. With minimal modifications, the ancillary knowledge needed to interpret a principle is thus the same as that outlined in Table 9.1 for any property concept.

Thus, Table 9.1 when applied to a principle, asserts that the specification of the principle can be achieved by a formal summary description (such as an equation), by informal qualitative statements, or by a detailed procedure which specifies what must be done to determine that the specified principle is true. The specification of the *value* of a principle is trivial (i.e., this value is simply "true"). The specification of independent variables includes again the specification of basic independent variables which need be specified and the specification of the relevant properties thereof. (For example, in the case of Newton's motion principle ma = **F**, the basic independent variables are some specified *particle,* some *other particle* with which it interacts, some specified *time,* and some specified *inertial reference frame.* The relevant properties of these independent variables are the mass m of this article, its acceleration **a** at this time relative to the specified reference frame, and the force **F** on this particle by all other particles interacting with it.) These remarks should suffice to indicate that our entire previous discussion concerning concepts is equally applicable to principles relating previously defined concepts.

IMPLICATIONS FOR LEARNING OR TEACHING

The preceding sections have sought to identify and explicate the ancillary knowledge required to interpret scientific concepts or principles. The discussion has made apparent that this ancillary knowledge is quite large and extends considerably beyond mere definitions of concepts or statements of principles. Such knowledge is commonly possessed by any expert, although he or she may not be consciously aware of its existence or able to articulate it explicitly. On the other hand, the acquisition of such knowledge by students is a demanding task.

The following paragraphs outline briefly the difficulties faced by students trying to learn unfamiliar concepts or principles. Then they explore the prospects of instructional methods exploiting the analysis of the preceding sections to teach concepts and principles more effectively.

LEARNING DIFFICULTIES

Anyone trying to learn an unfamiliar scientific concept or principle faces appreciable difficulties. Some of these are due to intrinsic characteristics of such scientific concepts or principles: (1) As discussed in the preceding sections, the knowledge required to interpret and apply such a concept or principle is considerable and sometimes subtle. (2) This knowledge often demands meticulous attention to details and requires fine discriminations to achieve the unambiguities required for accurate scientific predictions.

Other difficulties are characteristic of the person in the role of student trying to learn new concepts or principles: (1) A student brings to a learning situation many concepts and principles acquired in daily life or from more formal prior learning experiences. Hence the student's preexisting knowledge must be appropriately modified or transcended before new concepts or principles can be used without confusion and integrated into a new knowledge structure. (2) A student, unless thoroughly versed in scientific thinking, approaches learning from the vantage point of daily life where concepts or principles are adequately useful even if they are specified vaguely and somewhat inconsistently. Hence everyday concepts (e.g., "chair," "color," etc.) are often adequately specified by reference to prototypical cases which can be readily recognized or used for approximate comparisons. By contrast, scientific concepts need to be specified by explicit rules to ensure that they have unambiguous meanings. The learning of scientific concepts is thus a demanding task, rather different from the learning of concepts in daily life, and is correspondingly quite difficult for novice students unfamiliar with this mode of learning.

How effective are common teaching methods in dealing with these learning difficulties? Methods commonly used to teach concepts or principles involve presenting a new concept or principle, exemplifying the concept or principle in some special cases, and then providing students with practice in applying the concept or principle in various situations. Through a process of trial-and-error learning, students then gradually learn to avoid mistakes and to use the concept or principle more reliably.

There is considerable evidence that such teaching methods are neither very efficient nor effective. Indeed, after formal instruction and after months (or even years) of using a scientific concept or principle, many students still exhibit gross misconceptions, confusions, and other persistent errors (e.g., McCloskey, Caramazza, & Green, 1980; Trowbridge & McDermott, 1981). Furthermore, although students may *nominally* be familiar with certain concepts or principles, they often do not feel comfortable to use them spontaneously as intellectual tools facilitating their own thinking.

TEACHING APPLICATIONS

The analysis in the preceding sections identifies various kinds of important ancillary knowledge required to make a concept or principle effectively usable. This analysis can be used as the basis for instructional methods which teach such ancillary knowledge explicitly. It can also help to diagnose the causes of students' observed errors and difficulties.

The following paragraphs outline some suggested teaching methods based on this analysis. Although these suggestions are tentative and based on limited evidence, they provide a systematic approach suitable for further study and improvement.

Teaching Particular Concepts

A very common instructional aim is to teach students *particular* scientific concepts or principles (e.g., particular concepts such as "acceleration"). The ancillary knowledge summarized in Table 9.1 can then be used by an instructor, textbook, or other instructional medium to make explicit the ancillary knowledge required to interpret the particular concept of interest. (For example, the instructor can identify what particular independent variables are necessary to specify fully the concept "acceleration"; or the instructor can identify the likely error caused by confusion between the concept "acceleration" and the concept "velocity.") Systematic instruction then involves teaching students explicitly these specific kinds of ancillary knowledge at the time when the unfamiliar concept is first encountered. Indeed, the entries listed in Table 9.1 can easily be converted into specific *questions* which any student should be able to answer about the particular concept (e.g., questions such as "what is the procedure used to specify the meaning of the concept acceleration?").

Not only must one ensure that students display explicit familiarity with the various kinds of ancillary knowledge about a concept, but also that they actually *use* this knowledge when applying a concept. (For example, students should spontaneously answer questions about the acceleration by applying the *procedure* used to define this concept.) It is advisable that students acquire and consolidate this ancillary knowledge about a concept in the context of relatively simple questions and exercises. Only afterwards should they be asked to apply the concept in more complex problems.

Effective use of a concept requires that the ancillary knowledge about the concept become ultimately intuitive and habitually used. Needless to say, this requires adequate practice, but the *right kind* of practice specifically suggested by the analysis of the concept. Furthermore, explicit awareness of this ancillary knowledge can be useful to students, even after a concept has become intuitively familiar, since such explicit knowledge helps to debug errors or to cope with novel situations.

I have recently tried to exploit some of these teaching guidelines in actual classroom situations. This experience indicates that explicit teaching approaches based on the analysis in this paper can be very useful in practice. For example, it is very helpful to ask students to verbalize and apply *procedures* for identifying concepts. It also helps avoid many confusions to insist that students use full verbal expressions (such as "force *on* what *by* what"). However, the implementation of teaching procedures based on such an explicit analysis reveals also particularly clearly some general issues and difficulties inherent in any teaching process, issues which are worthy of further study in their own right.

Teaching Conceptual Learning Skills

The preceding comments have dealt with the teaching of *particular* concepts or principles. A much more ambitious instructional goal would involve teaching students the *general skill* enabling them to learn effectively *any* newly encountered concept or principle. The analysis presented in the preceding pages, as summarized in Table 9.1 is again basic to the systematic teaching of such a general learning skill. But now students would have to be taught the *general* ancillary knowledge required to make *any* concept or principle effectively usable, and would themselves have to translate this general knowledge into specific knowledge about any particular concept. This is clearly a much more difficult teaching task, but one of great importance. Indeed, successful implementation of such instruction would make students better independent learners who know explicitly what they need to study to achieve competent use of any new concept.

There is evidence that such instruction can be successfully implemented in practice. For example, a few years ago some collaborators and myself (Reif, Brackett, and Larkin, 1976), using a rather rudimentary analysis of concept learning and some primitive teaching methods based on this analysis, were able to show that students could be taught to become significantly better independent learners of new concepts. The more extensive analysis presented in the preceding pages, together with more explicit teaching methods, promises to lead to much more effective teaching of such general conceptual learning skills.

REFERENCES

Clement, J. (1982). Students' preconceptions in introductory mechanics. *American Journal of Physics, 50,* 66–71.
diSessa, A. A. (1982). Unlearning Aristotelian physics: A study of knowledge-based learning. *Cognitive Science, 6,* 37–75.
Larkin, J. H., and Reif, F. (1979). Understanding and teaching problem solving in physics. *European Journal of Science Education, 1,* 191–203.
McCloskey, M., Caramazza, A., and Green, B. (1980). Curvilinear motion in the absence of external forces: Naive beliefs about the motion of objects. *Science, 210,* 1129–1141.
Reif, F. (1981). Teaching problem solving or other cognitive skills: A scientific approach. *The Physics Teacher, 19,* 310–316.
Reif, F., Brackett, G. C., and Larkin, J. H. (1976). Teaching general learning and problem solving skills. *American Journal of Physics, 44,* 212–217.

Reif, F., and Heller, J. I. (1982). Knowledge structure and problem solving in physics. *Educational Psychologist, 17*, 102–127.

Trowbridge, D. E., and McDermott, L. C. (1981). Investigation of student understanding of the concept of acceleration in one dimension. *American Journal of Physics, 49*, 242–253.

Viennot, L. (1979). Spontaneous reasoning in elementary dynamics. *European Journal of Science Education, 1*, 205–221.

10

THE ROLE OF INTELLECTUAL ENVIRONMENT IN THE ORIGIN OF CONCEPTIONS: AN EXPLORATORY STUDY

Mariana G. A'B. Hewson

During a lifetime, we acquire knowledge which influences the way we interact with others, with our environment, and even with our own cognitive systems. Norman (1981) suggests that a study of human knowledge (cognitive science) needs to incorporate a study of world knowledge (anthropology and sociology), that is, that the 'external' aspects of a functioning cognitive system also need to be investigated.

External aspects concern the influences of the intellectual (i.e., cultural knowledge and beliefs) and physical environment on cognition. Norman (1981, p. 278) describes cultural knowledge as a "subset of general knowledge that is passed on from generation to generation" by society. The physical environment concerns the facts and events of the real world which provide a constant input, interacting with the cultural knowledge. Indeed, Norman (1981) suggests that the way in which individuals perceive their physical environment is affected by or even altered by their cultural knowledge. This idea is similar to that of Petrie (1976) and Toulmin (1972). Petrie (1976) describes individuals as having 'representational schemes.' These are seen as existing in a dynamic interaction between the cultural and social beliefs of the society, the prevailing paradigms or theories of the society, and the input from the facts and events of the physical world. This interpretationalist view of concepts espoused by Petrie (1976) is similar to Toulmin's (1972) idea of conceptual ecology. Both see thought in a relativistic perspective and interpret concept formation according to the varied mental sets of individuals which are a function of their intellectual and physical environment.

CONCEPTUAL ECOLOGY

The structure and development of knowledge can be viewed in terms of the biological metaphor of ecology in which people's ideas or concepts exist as a result of a process of natural selection (Toulmin, 1972). The intellectual environment in which a person lives (including cultural beliefs, language, accepted theories, as well as observed facts and events) favors the development of some concepts and inhibits the development of others. Thus, the intellectual environment acts as an ecological niche. Conceptual ecology involves a dynamic interaction between a person's knowledge structures and the intellectual environment in which he or she lives.

According to Toulmin (1972), concepts are grouped into "conceptual frameworks" which serve to predict and explain facts and events. One such conceptual framework, Newton's theory of mechanics, is thus, in Toulmin's terms, an intellectual adaptation to the ecological niche of the scientific knowledge and endeavors of the times. Toulmin then draws the implication that in different historical and cultural conceptual ecologies different conceptual frameworks are likely to evolve in order to explain the same natural phenomena. These "alternative frameworks" are not necessarily misconceptions, but are cultural adaptations of concepts and procedures to the specific demands of particular intellectual niches.

In this chapter, alternative frameworks are referred to as "alternative conceptions." The term *conception* is used to indicate a functional unit of thought which has both propositional (knowing that) and procedural (knowing how to) aspects (Shavelson, 1974).

ALTERNATIVE CONCEPTIONS

There is a growing body of research evidence to support the claim of the existence of alternative conceptions, particularly in scientific subjects. Driver and Easley (1978) have reviewed early research in this area and Driver and Erickson (1983) have a comprehensive survey of more recent research activity in this field. They show that, researchers from a number of different countries have been working on student alternative conceptions in subjects as diverse as mechanics, dynamics, heat and temperature, electricity, energy, potential, pressure, gravity, vectors, particulate theory, the earth as a cosmis body, mass, volume and density, evolution, heredity, and the circulatory system. It should be noted, however, that the existence of alternative conceptions in childrens' thinking was documented as long ago as the 1920's by Piaget (1929). His meticulous and systematic use of the open interview is responsible for much of the current research in cognitive structure and conceptual change.

The occurrence of alternative conceptions is consistent with the constructivist view of learning which characterises the learner as an active participant in the construction of his or her own knowledge. This view carries the implication that different people strive to make sense of the world; that they use their idiosyncratic existing knowledge to do this and therefore different people will acquire different conceptions even when presented with the same information. In this way, it is possible for different people to construct alternative conceptions from the same information.

Up to the present, most of the research concerning alternative conceptions has been naturalistic in form. In other words, researchers have been mainly concerned with the problems of eliciting and analyzing students' knowledge and documenting their alternative conceptions in a variety of subject areas. A few researchers have attempted to investigate the role of alternative conceptions in learning, and attempts have been made to improve instructional strategies which deal with the alternative conceptions (Champagne, Klopfer, & Gunstone, 1982; M. G. Hewson, 1982; M. G. Hewson & P. W. Hewson, 1983; Nussbaum & Novick, 1981).

This chapter focuses on the theoretical issue concerning the origin of alternative conceptions in the context of peoples' intellectual environment. Driver and Erickson (1983) suggest that kinaesthetics or sense experiences, language and available metaphor, and analogic reasoning based on perceptual similarities between new and prior experiences, are useful areas of investigation. In this chapter, the formation of conceptions is discussed in the context of the semantic metaphors which exist in both cultural knowledge and the prevailing theoretical paradigms of a group of people.

THE ROLE OF METAPHOR IN INTELLECTUAL ENVIRONMENT

Lakoff and Johnson (1980) suggest that fundamental concepts emerge from experience, and that every experience takes place in the context of cultural assumptions. In other words, they argue, "truth is always relative to a conceptual system, that any human conceptual system is mostly metaphorical in nature, and that therefore there is no fully objective, unconditional, or absolute truth" (p. 185). Furthermore, the social reality defined by a culture affects its conception of facts and events in the world at large. Abstract concepts are discussed by Lakoff and Johnson (1980) as systems of related metaphors, which arise naturally from physical and cultural experience.

When basic metaphors which are implicit in a scientific theory are extensions of basic metaphors in our everyday conceptual frameworks, the theory
is experienced as "intuitive" or "natural." When, however, there is a disparity between the metaphors in scientific and the everyday conceptual frameworks the theory may be experienced as "counter-intuitive" or "unnatural."

The argument put forward by Lakoff and Johnson (1980) is entirely compatible with the notion of conceptual ecology (Toulmin, 1972). In the latter
view, the conceptions of the individual exist in a dynamic interaction with
the various aspects of intellectual environment, while the former view holds
that metaphorical concepts are grounded in experience and that they in turn
influence the way in which everyday experiences are perceived.

The history of science shows a number of shifts in paradigms. An example
is found in the change from Newtonian to Einsteinian physics. Sutton (1980)
suggests that the established tradition in science deals with denotative
meanings (rigorous definition) while everyday experiences are seen in terms
of connotative meanings (the framework of associations and implications),
and that there is ample evidence of shifts in meaning over the years. Sutton
describes the effort of Boyle to reject the vague generalized meanings associated with the discussion of the four elements, earth, air, fire, and water, and
to redefine an element as a "perfectly unmingled substance." Further shifts
can be recognized in the use of the terms 'phlogiston' in burning and other
oxidation phenomena, and 'caloric' in phenomena involving the transmitting
of heat energy. These outdated notions involved metaphors. Sutton's (1980)
example of the old use of the word *sail* for sailing ships being extended for
describing the motion of steamships as "sailing away" involves using the
word metaphorically. In much the same way, the word 'caloric' is a metaphor implying a fluid-like substance which enters objects, thereby making
them hot. This metaphor is probably rooted in experiences with the phenomenon of the expansion of hot bodies and was provided with denotative
meaning by Lavoisier (1789). Sutton (1980) suggests that the evolution of
scientific vocabulary often involves extensions of meaning through metaphorical reapplication of existing words. Similarly, Lakoff and Johnson (1980)
suggest that much of what goes on in our world is given meaning by metaphorical concepts, which may themselves change in different cultural environments, or over time.

If metaphors can be seen as a means of expression, then it is possible that
both denotative and connotative meanings are composed of metaphors, and
that metaphors are in fact a large component of what has been described in
this chapter as "intellectual environment." The question then is whether
particular conceptions (either scientific or alternative) concerning natural
phenomena can be attributed to metaphors which may exist in either denotative or connotative meanings.

EMPIRICAL EVIDENCE FOR THE INFLUENCE OF METAPHORS

In an exploratory study, Hewson and Hamlyn (1983) attempted to establish a logical fit between the intellectual environment of a specific group of people and their conceptions of heat. This involved establishing what conceptions the subjects used when explaining tasks concerning phenomena associated with heat, and then discussing these in the context of known prevailing metaphors in the subjects' everyday life. In addition, this same approach was used to throw light on the findings of other researchers, such as Erickson (1979) who have been interested in the subject of childrens' conceptions of heat.

The study focused on African people in southern Africa who are collectively called the Sotho group. Setswana and North Sotho languages were predominant. The respondents included 10 schoolchildren (Grades 9 and 10), four unschooled adult workers, and six semi-schooled adult workers (Grades 7 to 10). All the subjects had an adequate knowledge of English and were living in urban or semi-urban areas at the time of testing.

The subjects in this study lived in the hot, arid area in the interior of southern Africa. The land has few natural agricultural resources and an adequate supply of water is a major concern of the people. This harsh environment appears to have given rise to a powerful metaphor concerning heat which pervades many aspects of life. Anthropologists (Hammond-Tooke, 1981; Schapera, 1979; Verryn, 1981) document that the Sotho group of peoples living in this area believe that "coolness is good" (implying health and social harmony) and the converse "hot is bad" (implying sickness and social disharmony). These ideas can be termed metaphors, and, as Lakoff and Johnson (1981) point out, metaphors are often grounded in experience. In this case, the metaphor appears to be grounded in the hardship caused by a hot, dry environment (Hammond-Tooke, 1981).

The heat metaphor is used in many instances in everyday life, such as birth, pregnancy, menstruation, death, and sickness; and a variety of situations involving negative feelings, for example, anger, impatience, and anxiety. A person who is in one of these situations is said to be 'hot.' Such a person's blood is thought to be particularly susceptible to heat, in which condition it is said to be "agitated" or "fighting with itself" (Schapera, 1979).

The formal Western scientific understanding of heat has undergone a paradigm shift. In the sixteenth and seventeenth centuries, scientists such as Lavoisier believed in the caloric view of heat, in which heat was believed to be a kind of fluid called "caloric" which flowed into or out of bodies when they became hot or cold respectively (Gordon, Neser, Pienaar, & Walters, 1970; Zemansky, 1957). This idea can be described as a metaphor which seemed to give a certain amount of explanatory power regarding the phenomenon of heat. The concept of heat is difficult because it is abstract and only the effects of heat are visible and concrete. While the metaphorical concept arises naturally from physical experiences, the metaphor serves to give only partial meaning to the abstract concept. In this case, the metaphor of heat as a caloric had only limited usefulness, and in the nineteenth century, the kinetic view of heat became prevalent. According to this latter view, heat is one form of energy in transit. In other words, it is energy which is transmitted from an object at a high temperature to another at a low temperature (Brink & Jones, 1977; Zemansky, 1957). The kinetic theory of heat is taught in schools and it is clear that the notion of heat as "heat energy" is quite commonly accepted at the junior high school level but would not necessarily be accepted at university, where heat is seen strictly as a flow of energy.

The M. G. Hewson and Hamlyn (1984) research aimed to establish whether the particular subjects did indeed subscribe to the heat metaphor. This was done by using a semi-structured interview called Interviews-about-Instances (See Osborne & Gilbert, 1980). A variety of life situations were protrayed by means of photographs and a verbal description of what was happening accompanied each photograph. The person was asked, "Would you say this person is 'hot' in this situation?" and, "If so, why would you say that?". The results of these interviews showed that 16 of the 20 respondents agreed that at least some of the instances would involve a person being 'hot' in the metaphorical sense. This suggests that the heat metaphor is still commonly used by Sotho's who are urban, somewhat acculturated to Western life and thought, and who are able to converse in English as a second language.

The second part of the research aimed to ascertain what conceptions of heat were used by these people to explain a variety of situations involving heat phenomena. This part of the research mainly replicated the experimental procedure of Erickson (1975), and made use of unstructured interviews which focused on five tasks. These dealt with the phenomena of hot substances, expansion and contraction, and conduction. The recorded interviews were analyzed using techniques developed by Erickson (1975) and M. G. Hewson (1982), and involved the development of a Conceptual Profile Inventory. This instrument showed tne range of conceptions used by the subjects, as well as the number of subjects who subscribed to each conception. The data showed that the most common alternative conception was one in which the subjects described the heat phenomena at a micro level in terms of physical phenomena visible at a macro level, such as boiling, burning, or melting. Of the 20 respondents, 17 used this alternative conception. Other alternative conceptions such as "particles split or multiply when an object is hot, and recombine when an object is cold" (7 respondents), "particles expand when an object is hot and contract when cold" (8 respondents) were documented. The alternative conception that "heat is a caloric" had the lowest occurrence (only 2 of the 20 respondents). On the other hand, a surprising result was the large number of people (12 respondents) who subscribed to the scientific category of prekinetic/kinetic conceptions. While it was true that these responses came mainly from the semi-schooled workers and schoolchildren, it was found that many of these subjects also subscribed to the metaphorical heat conceptions. These results show a divergence from those found by researchers in Western environments. Albert (1978), Erickson (1975, 1979, 1980), Tiberghien (1980), and Shayer and Wylam (1981) all report that caloric conceptions of heat are relatively common in school children.[1] Why should this be so? A tentative claim made by M. G. Hewson and Hamlyn (1984) is that the prevailing metaphors in people's intellectual environments may indeed produce cognitive differences. The Sotho group, who generally subscribed to the cultural heat metaphor, showed a relatively high number of responses which were described as prekinetic/kinetic and a low number of caloric responses. This suggests that the Sotho language may predispose speakers toward a meaning of heat which has connotative meanings influenced in some way by the powerful heat metaphor in Sotho cultural beliefs. This metaphor which concerns the conception of 'hot' being a personal condition involving "agitated or disturbed" blood, fits logically with the prekinetic views expressed by many Sotho subjects. Conversely, the findings by Western researchers mentioned earlier, that caloric conceptions of heat are commonly found in schoolchildren, suggest that Western languages may have a predisposition towards a meaning of heat which has caloric connotations. This predisposition would be caused by the vestiges of the early caloric metaphor that have given rise to words such as *heat* itself, and also *heat flow*, 'heat capacity,' and 'calorie.' Harris (1981) claims that Western students use caloric conceptions because the European languages such as English convey the wrong idea by using words for heat which are outdated and act as red herrings. This would be true of the everyday usage of the word 'heat,' for example: 'Is there any heat in that pot of tea?' or, 'The heat went out of the sun.' The source of confusion, therefore, is a semantic one, which may be traced to an outdated metaphor, which has become embedded in the theoretical paradigms and cultural beliefs of Western culture.

[1]While the Hewson and Hamlyn research methodology followed closely the clinical interview style proposed by Piaget (1929) and replicated some of Erickson's (1975) experiments, their results are not strictly comparable, in a statistical sense, with those of independent Western researchers. The inferences drawn, therefore, remain tentative.

CONCLUSION

The M. G. Hewson and Hamlyn (1984) research provides an indication of the effect of aspects of the intellectual environment on concept acquisition. In this case, the authors suggest that linguistic metaphors which are rooted in the history and culture of a people influence, at least to some extent, the way in which Sotho people explain one particular natural phenomenon, namely, heat. The role of the intellectual environment is clearly a factor in shaping cognitive structure. The difficulty is to show how this is so. Although the M. G. Hewson and Hamlyn (1983) study was exploratory, it suggests a fruitful line of research concerning the effect of cultural metaphors on the origin of conceptions.

REFERENCES

Albert, E. (1978). Development of concepts of heat in children. *Science Education*, 389–399.

Brink, B. du P., & Jones, R. C. (1977). *Physical Science 8.* Cape Town: Juta and Co., Ltd.

Champagne, A. B., Klopfer, L. E., & Gunstone, R. F. (1982). Cognitive research and the design of science education. *Educational Psychologist, 17*, (1), 31–55.

Driver, R., & Easley, J. (1978). Pupils and paradigms: A review of literature related to concept development in adolescent science students. *Studies in Science Education, 5*, 61–84.

Driver, R., & Erickson, G. (1983). Theories in action: Some theoretical and empirical issues in the study of students' conceptual frameworks in science. *Studies in Science Education, 10*, 37–60.

Erickson, G. (1975). *An analysis of childrens' ideas of heat phenomena.* Unpublished doctoral dissertation, University of British Columbia, Vancouver.

Erickson, G. (1979). Childrens' conceptions of heat and temperature. *Science Education, 63*(2), 221–230.

Erickson, G. (1980). Childrens' conceptions of heat. *Science Education, 64*, 323–338.

Gordon, W., Neser, G. O., Pienaar, H. N., & Walters, S. W. (1970). *Basic senior physical science.* Cape Town: Maskew Miller Ltd.

Hammond-Tooke, W. D. (1981). *Patrolling the herms: Social structure, cosmology, and pollution concepts in Southern Africa.* Raymond Dart Lecture No. 18. Johannesburg: University of the Witwatersrand Press.

Harris, W. F. (1981). Heat in undergraduate education, or isn't it time we abandoned the theory of caloric? *International Journal of Mechanical Engineering Education*, 9(4), 317–321.

Hewson, M. G. A'B. (1982). Students' existing knowledge as a factor influencing the acquisition of scientific knowledge. Unpublished doctoral dissertation, University of the Witwatersrand, Johannesburg.

Hewson, M. G. A'B. & Hamlyn, D. (1984). The influence of intellectual environment on conceptions of heat. Paper presented at the annual meeting of the American Educational Research Association, Montreal. *European Journal of Science Education*, 6(3), 245–262.

Hewson, M. G. A'B., & Hewson, P. W. (1983). The effect of instruction using students' prior knowledge and conceptual change strategies. *Journal of Research in Science Teaching*, 20(8), 731–742.

Lakoff, G., & Johnson, M. (1980). *Metaphors we live by.* Chicago: University of Chicago Press.

Lavoisier, A. L. (1982). *Memoir on heat: Read to the Royal Academy of Sciences, 28 June, 1783* (H. Guerlac, Trans.). New York: Watson.

Norman, D. A. (1981). Twelve issues for cognitive science. In D. A. Norman (Ed.), *Perspectives on cognitive science*. New Jersey: Ablex Publishing Co.

Nussbaum, J., & Novick, S. (1981). Brainstorming in the classroom to invent a model: A case study. *School Science Review, 62*(221), 771–778.

Osborne, R. J., & Gilbert, J. K. (1980). A method for investigating concept understanding in science. *European Journal of Science Education, 2*(3), 311–321.

Petrie, H. G. (1976). Evolutionary rationality: Or can learning theory survive the jungle of conceptual change? *Philosophy of Education*, 117–132.

Piaget, J. (1929). *The Child's Conception of the World*. London: Routledge and Kegan Paul.

Schapera, I. (1979). Kgatla notions of ritual impurity. *African Studies, 38*, 3–15.

Shavelson, R. J. (1974). Methods for examining representations of a subject matter structure in students' memory. *Journal of Research in Science Teaching, 11*(3), 231–250.

Shayer, M., & Wylam, H. (1981). The development of concepts of heat and temperature in 12–13 year-olds. *Journal of Research in Science Teaching, 18*(5), 419–434.

Sutton, C. (1980). Science, language and meaning. *The School Science Review, 62*, 47–56.

Tiberghien, A. (1980). Modes and conditions of learning—an example: The learning of some concepts of heat. In W. F. Archenhold, R. H. Driver, A. Orton, & C. Wood-Robinson (Eds.), *Cognitive development and research in science and mathematics*. Leeds: University of Leeds Printing Service.

Toulmin, S. (1972). *Human understanding: Vol. 1. The Collective use and evolution of concepts*. Princeton: University of Princeton Press.

Verryn, T. (1981). 'Coolness' and 'Heat' among the Sotho peoples. *Religion in Southern Africa, 2*, 11–38.

Zemansky, M. W. (1957). *Heat and thermodynamics* (4th ed.). New York: McGraw-Hill.

11

EFFECTING CHANGES IN COGNITIVE STRUCTURES AMONG PHYSICS STUDENTS

Audrey B. Champagne, Richard F. Gunstone, and
Leopold E. Klopfer

*Children and adults often express quasi-Aristotelian dynamics concepts re-
sembling momentum, acceleration, and the vectorial decomposition of gravitational attraction. Unfortunately, most curricula in physics suggest a purely
descriptive (kinematical) investigation of position and time, and provide no
means for people to test their own theories of dynamics and to restructure
them.*

(Easley, 1971, p. 156)

INTRODUCTION

It is now well established that, before receiving formal physics instruction,
students possess knowledge which leads them to idiosyncratic interpretations of real world events. Physicists, however, see the same events as
exemplifying principles of physics. Because students' interpretations are
often at variance with physicists' interpretations, they have been labeled
misconceptions, alternative frameworks, personal models of reality, etc. Although forms of such idiosyncratic knowledge have been observed in various
content areas, elementary mechanics has been the focus of a number of
studies. Our own work has established the common existence of knowledge
that is logically antagonistic to the tenets of physics both among students
beginning a formal study of mechanics (Champagne, Klopfer, Solomon, &
Cahn, 1980; Gunstone, Champagne, & Klopfer, 1981) and among students
who have successfully completed an introductory physics course (Champagne, Klopfer, & Anderson, 1980; Gunstone & White, 1981), and we have
reviewed the studies that explore the nature of pre-instructional mechanics
knowledge and its structural organization (Champagne, Klopfer, &
Gunstone, 1982). As the number of studies of pre-instructional knowledge of
mechanics has grown, it has become clear that naive schemata students use

to interpret the motion of objects are extremely resistant to change during formal instruction. As a consequence, there has been considerable interest in attempts to design instruction to facilitate cognitive structure change. In considering this instructional problem, we have previously proposed four dialogue-based instructional strategies specifically designed to alter declarative knowledge about the motion of objects that the students bring to instruction in mechanics (see Chapter 5, this volume). Here, we report on our investigations of the extent to which one of the strategies facilitates change in two quite different groups—middle school students designated as academically gifted and with a demonstrated interest in science in Pittsburgh, and non-physics-major university science graduates studying to become high school science teachers in Victoria—and discuss particular features of the strategy that may be of significance in promoting change. This chapter begins with a brief description of the structure of the instruction and the probes used to gather information about aspects of students' cognitive structures. Some data are then presented and some preliminary conclusions drawn.

INSTRUCTION

The instruction was designed to provide considerable opportunity for students to argue their own interpretations of both specific events and general relationships. It began with a substantial discussion of the question, "What are some of the things we can say about the relationships between force and motion?" and subsequently explored several pertinent topics, including methods of describing motion and force, falling bodies, motion on inclined planes, and forces on objects in a variety of contexts. Demonstrations and direct hands-on experience were used. Group sizes were small and the time spent with each group reasonably long. (Details for each group are given below.) Because of the considerable differences in the two groups, there were necessarily differences in the content of the instruction; however, the basic instructional strategy—ideational confrontation—used with both groups was the same.

Briefly, this strategy first asks the students to be explicit about the notions they use to explain or make predictions about a common physical situation— say, the motion of an air-filled balloon as the air rushes out of it, or the motion of two sleds, one empty and the other loaded, that are released simultaneously at the top of an ice-covered ramp. After a physical situation is described, each student develops an analysis that supports his or her prediction, and then individual students present their analyses to the class. Inevitably, controversies arise, and analyses are modified. Typically, students with different interpretations begin to attempt to convince others of the validity of their ideas. As a student or group of students defends a position, concepts become better defined, and underlying assumptions are stated explicitly. The net result is that each student is explicitly aware of his or her existing notions about motion that were used in the analysis of the situation of interest. Typically, as a result of these discussions, students become dissatisfied with their current theories.

At this point in the ideational-confrontation strategy, the instructor demonstrates the physical situation (say, he or she lets the air escape from the balloon or lets the two sleds slide down the icy ramp) and presents a theoretical explanation of the results, using the particular science concepts, principles, and theory which the demonstration illustrates. The concepts, principles, and theory in the instructor's scientific explanation usually are different in significant respects from the notions the students used in their analyses. In further discussions, the students compare the elements of their analyses of the situation with the scientific analysis given by the instructor, and identify similarities and differences. This exercise requires the students to confront inconsistencies between their existing notions and the content of the science instruction. The ideational confrontation enables each student to become aware of his or her existing notions and of the need to reconcile them with the science concepts and principles that are to be learned.

PROBES OF COGNITIVE STRUCTURE

Assessments were made of students' cognitive structures before and after the instruction by administering a variety of probes. Detailed discussions of the possible methodologies which could have been employed and why the particular ones were chosen is discussed elsewhere (Champagne, Hoz, & Klopfer, 1984). Descriptions and some brief comments about the several probes follow.

Five probes of cognitive structure were used. Three of these have previously been used in psychometrically oriented studies of cognitive structure; the other two have been used previously in studies of students' conceptions of the motion of objects. The first three probes were group-administered, while the last two were administered individually. Cognitive structures derived from analyses of data collected with four of these probes describes a non-contextual structural organization of concepts. Although the structures obtained by these measures are in one sense contextual because the students are well aware that the content is science or physics, they are non-contextual in the sense that the concepts are not being applied to the solution of a specific task, for example, the comprehension of text or the solution of a numerical or qualitative problem. An issue is the consistency of cognitive structures obtained in different contexts. That is, are the structures the same when the task is, say, to "arrange these terms as you think about them" as when the task is to analyze the motion of an object on an inclined plane. Thus, with the fifth probe, data on the structural organization of the target concepts in a specific context and students' application of the concepts were obtained by presenting the students with a physical situation, describing a manipulation to be performed, and asking them to make a prediction about the outcome of the manipulation and to describe the knowledge (facts and principles) that they used to make the prediction. Concepts the students did not use spontaneously were introduced by asking the students to evaluate predictions and explanations of the same situation given by other students.

FREE SORT TASK

The free sort task is a concept classification and categorization task that required students to categorize 17 physics concepts: acceleration, change of motion, direction, displacement, distance, force, frame of reference, inertia, mass, magnitude, motion, position, relative, resultant, speed, time, and velocity. No criteria for categorization were suggested. Various versions of free sort tasks have been used previously, (e.g., Gorodetsky & Hoz, 1985; Miller, 1969; Shavelson & Stanton, 1975), and Wiley (1967) proposed a multivariate scaling technique, latent partition analysis, for analyzing free sort task data. In our version of the task, student instructions began with an example of two ways in which 7 non-physics terms could reasonably be classified in order to emphasize the absence of a single, correct answer to the task. The instructions then reiterated the absence of a right answer, indicated that the words to be classified were on cards inside an envelope provided, asked that the words be classified into "categories that include words that you think belong together," and pointed out that any number of categories was possible and that any category could contain any number of words. On completion of the card categorizing, students transferred their grouping to an answer sheet containing 12 boxes. No time limit was applied to the task.

TREE CONSTRUCTION TASK

In the tree construction task, the students were asked to construct a linear undirected graph (tree) of the same 17 physics concepts used in the free sort task. This type of task was devised by Rapoport (1967; Fillenbaum & Rapoport, 1971) as a method of measuring semantic distance between concepts. Hierarchical clustering (S. C. Johnson, 1967) is the scaling method that has been most commonly employed in the analysis of Tree Construction Task data. In our version of the task, the students were given an alphabetical list of the 17 concepts and were instructed to "pick the two words from the list which you think are *most related* to each other," write the words in the center of the page, and connect them with a line numbered "1." This was then built on by adding another word from the list which was seen as "most related" to one of the two already used, then connecting with a line numbered "2." Either this procedure continued or a new tree was started, depending on the student's judgment of the extent of relatedness. Instructions required students to eventually join up trees if they had produced more than one so that the total response to the task comprised the 17 words and 16 numbered connections. Again, no time limit was applied.

WORD ASSOCIATION TASK

The word association task gave the student a one-minute time period to generate as many free associations as possible when each of these 5 concepts was successfully used as a stimulus: force, mass, speed, inertia, change of motion. This type of task derives from work on the determination of the associative meaning of concepts (e.g., Deese, 1965), and it has been used in several studies concerned with the cognitive structures of physics students (e.g., P. E. Johnson, 1965; Preece, 1976; Shavelson, 1972). In our word association task, when the students had completed each set of associations to a stimulus, they were asked to compose a sentence for each individual association which contains both the stimulus word and the associated word. The sentence-composing procedure, previously used by Gunstone (1981), is intended to elicit information about the nature of the associative link, and this part of the task is given without a time limit.

CONCEPT STRUCTURING ANALYSIS TECHNIQUE

The concept structuring analysis technique (ConSAT) task used the same 17 concepts that appeared in the free sort and tree construction tasks. The ConSAT was devised by Champagne and Klopfer as a way of eliciting graphic representations of students' cognitive structure (Champagne, Klopfer, De Sena, & Squires, 1981). For our student interviews using the ConSAT, the concepts were written on small cards, and the student was asked to identify which ones he or she recognized, and then to provide a definition for the recognized concepts. Finally, the student was asked to arrange the recognized concepts on a sheet of paper "in a way that represents how you think about the words and about the relationship between them." When the arrangement was completed, the grouped concepts were circled and the student was asked why he or she made this arrangement and to explain particular groupings. The student's answers were tape-recorded and written on the sheet containing the arrangement.

DOE INTERVIEW

For this probe of the student's cognitive structure, the demonstrate, observe, explain (DOE) task, described in the review of our earlier research (see Chapter 5, this volume) was modified for use in an individual interview context and to elicit more extensive responses from the student. For a sequence of physical demonstrations, the student was asked to give a prediction of the outcome of the demonstration, to explain the basis of the prediction (knowledge used, relevant personal experience, etc.), to observe the demonstration, and to consider any differences between prediction and observation. In addition, prior to observing the demonstration, the student considered a set of predictions and explanations given on previous occasions by other students and was asked to describe the way in which various concepts were being used in these students' answers and to compare them with his or her own ideas. The entire interview was tape-recorded.

PRE- AND POST-TESTS

Identical task forms were used before and after the instructional period with the exception of the DOE interview. On the pre-test, the DOE interview involved a sequence of demonstrations based on a dime placed on a piece of cardboard that in turn could move along a suspended, sloping meter rule. On the post-test, the demonstrations were based on a dime placed on a block of wood which was in turn placed on an inclined plane. That is, the demonstrations in the two DOE interviews had different surface features and the same deep structure.

THE PITTSBURGH GROUP

The instructional group was comprised of 13 middle school students selected from school districts in Allegheny County. These students were all designated as academically gifted by the criteria set forth by the Pennsylvania Department of Education and all had demonstrated an interest in science out of school. A control group of 10 was selected from the same school districts by use of the same criteria. This group was also given the cognitive structure probes on two occasions separated by eight weeks, but did not experience any special program in this time.

The instructional period occupied 10 weeks, during which the students came to the Learning Research and Development Center at the University of Pittsburgh for one full day, from 9:00 A.M. to 2:30 P.M., each week. The first session was occupied primarily with an orientation to the program, organizational matters, and the pre-instructional student assessment. Most of the tenth session was devoted to presentations given by the students for their parents and regular science teachers. The post-instructional testing took place near the end of the ninth session. Hence, with allowances for breaks and lunch, the total instructional time available was approximately 30 hours. In each day's session, generally about 2 hours and sometimes more were spent in discussions of students' ideas and various issues concerning motion. These were the occasions when the ideational confrontation strategy was most directly applied. The rest of the time in each session saw the students engaged in carrying out various laboratory activities involving inclined planes, force tables, an air track, and other apparatus, or in performing computer-simulated mechanics experiments, or both. While the students worked on these activities in small groups, an instructor was generally at each station to raise questions and promote further discussions.

For the instructed group students as a whole, their previous formal instruction related to the physics of motion was minimal. The 6 eighth graders in the group had been exposed prior to the beginning of this study to the 3- to 4-week unit on motion, energy, and machines typically found in eighth-grade physical science or general science textbooks. The group's 6 seventh graders and 1 sixth grader had not received even that much formal instruction relating to mechanics prior to this study. Their in-school exposure to mechanics was limited to whatever may have been included in their elementary school science programs, few of which include very much. Two of the students reported that they had studied topics in mechanics quite extensively on their own.

FINDINGS FROM THE COGNITIVE STRUCTURE PROBES

In the tree construction task, the students were asked to select the pair of concepts they considered most related, the second-most-related pair and so on. We found that their choices for the early positions focused on a very limited number of the concept pairs. Of the 136 possible pairs, only 9 different pairs were cited on the pre-test as most related by one or more students, and only 7 different pairs were selected first on the post-test. All told, only 30 different pairs were selected first, second, or third by any student on the pre- and post-tests combined. Seven of these 30 pairs accounted for 54% of all the selections for the first three positions.

On the pre-test, the pairs most frequently selected as the most-related concepts were speed–velocity (three times) and motion–speed (three times). Solitary selection of most-related concepts went to each of these seven pairs: motion–change of motion, direction–distance, acceleration–velocity, inertia–speed, acceleration–speed, time–speed, mass–displacement. When the students repeated the tree construction task on the post-test, they again picked the speed–velocity pair most frequently (four times) as the most-related concepts, but now the force–motion pair was selected as the most related by three students and the acceleration–change of motion pair was selected for first position by two students. Pairs each selected once on the post-test as most-related were acceleration–force, distance–displacement, direction–distance, and magnitude–velocity. It is of interest to note that the most frequently selected first position pair on both the pre- and post-tests is one where the concepts (speed and velocity) are viewed as synonyms by the students. Also noteworthy is the observation that the concept of *force* does not figure in any of the most-related pairs at pre-test time, whereas it appears four times in first position pairs on the post-test. This change is consistent with one emphasis of the intervening instruction.

In the instructed group's responses to the word association task, the total number of associations made by the 13 students on the pre-test was essentially the same as on the post-test (290 vs. 292). However, the associations they made were very different. A total of 137 terms was used to generate the pre-test associations. After instruction, 179 terms were used to generate the associations. About 42% of the terms used on the pre-test were also used on the post-test. However, an analysis of the sentences generated using these repeated terms shows that, with a few notable exceptions, the relationship between the stimulus and the associated term changed from pre- to post-test.

Two instances where the relationship between the stimulus and associated term did not change were between force and motion (movement) and between mass and weight. In both instances, the maintained relationship was a scientifically incorrect one. Nonetheless, in most cases, the sentences containing both terms were not propositions directly relating to the concepts (e.g., "*mass* is not the same as *weight*"; "*force* is a *vector* quantity"), but rather sentences in which the two terms happened to co-exist (e.g., "The *force* of the *power* plant is unmanageable"). Most of the sentences were contextual (e.g., "I am *exerting* a *force* on this chair") and in no instance were concepts related in a formal way (e.g., "*Force* equals *mass* times *acceleration*"). In summary, two observations relevant to the relationships between terms are that there is little consistency between pre- and post-test associations and that there is no identifiable trend in the changes from pre- to post-test.

Before producing a concept map in the ConSAT Task, the students were asked to identify which of the 17 concepts they recognized and which they did not. On the pre-test, five different concepts were placed in the unrecognized pile by one or more students. The concepts identified as unrecognized were frame of reference (five times), resultant (four times), displacement (two times), velocity (one time), and relative (one time). Moreover, no definitions of any kind were recorded by 11 students for frame of reference, by 10 students for resultant, by 7 students for displacement, and by 2 students for relative. On the post-test, none of the 17 concepts was classified as unrecognized. Definitions were not given by 2 students for frame of reference and by 1 student for displacement.

The difference in knowledge of the concepts is also indicated by the pre- and post test difference in the terms for which the students gave definitions. On the pre-test, there were 38 places, involving 10 different concepts, where a definition was not even attempted. On the post-test, there were only 9 occasions, involving 6 concepts, where students did not give a definition. Approximately 75% of the 38 omitted definitions at pre-test were for the three concepts frame of reference, resultant, and displacement. At post-test, all of the terms were predominantly defined in a way which was consistent with a physical science definition, as opposed to merely an instance of the concept describing or providing an example which involves the concept. Also, at post-test the definitions incorporated markedly more physical science concepts from the list of 17 concepts itself. Changes also occurred in the structural organization of the terms. These changes point toward the development of increased concept differentiation. Both the pre- and post-ConSAT structures contain groups of terms. Each group has a superordinate term characterizing the terms in it and each of the groups is typically linked to another group by a relationship. Expert structures of the same terms do not have any groups—each term being linked to another by a specific relationship. We take this as evidence that the terms are not well differentiated by the students. Evidence that the students are beginning to differentiate the terms better is provided by the fact that in the post-instructional structures there are on average more groups and more explicit relations between individual terms within groups. This is in contrast with global statements about relationships between associated groups of terms that are evident in pre-instructional structures.

In the transcripts of the DOE interviews given before the instruction, the most striking feature of the students' analysis of the inclined plane is the absence of any of the target terms in the reasons the students give for their predictions. The students' analyses of the physical situation were consistently based on two surface features of the situation, the slope (angle of incline) of the plane, and the roughness of the surface on which the coin was placed. A few students mentioned friction and gravity as factors related to roughness and incline but no student attempted any analysis of the forces on the coin. On the post-test DOE interview, even though the students had been exposed to very specific instruction about the gravitational and frictional forces on objects in free fall, none of them applied anything resembling a force analysis to the coin on the block in the post-test problem. All used an argument based on surface features similar to that used on the pretest. Even though the information and skills necessary to attempt a force analysis were available to the students, the real-world nature of the problem triggered the well-practiced analysis.

THE VICTORIAN GROUP

The students involved in the Victorian part of the study were science graduates undertaking a one-year course in teacher training at Monash University with the aim of teaching high school biology or chemistry. In the Victorian context this means that they would also teach general or integrated science in years 7–10 of high school, and thus be involved in teaching physics concepts. As all of our work on instruction to change physics cognitive structures prior to this study had been with middle school students, a pilot program was run with eight trainee teachers. None of the data from this trial is reported here.

Because of course and administrative constraints, the group involved in the study were all volunteers. The choice to join the group was made on the basis of information describing what would take place as being "Physics for Non-physicists" directed toward developing an understanding of important ideas in elementary physics. Students had the option of joining this group or one of thirty or so others which ran across the total spectrum of issues of possible relevance to trainee teachers.

The group comprised six students. It is clear, given the circumstances described above, that all six both saw their understanding of physics to be inadequate and had considerable motivation to try to rectify this situation. The group met for two full days each week, from 9:30 A.M. to 4:00 P.M. The first day of the first and fourth weeks was given over to the group and individual tests of cognitive structure. Thus, the instruction occupied five days, or about 30 hours. A control group was generated, again by calling for volunteers. However, the resulting group differed considerably from the instructed group in terms of physics background, and consequently, the control group data are not considered here. This problem emphasizes that the instructed group cannot be considered a representative sample of the population of prospective secondary science teachers who have inadequate training in physics.

Because of the nature of the group, there were some differences in the conduct of the study by comparison with Pittsburgh. The Victorian students reacted individually (in writing) to the experiences of each day and to the whole sequence. In addition, on the first day they gave relevant biographical data and their perceptions of the areas of physics they believed they understood or did not understand. Discussions with the group tended to run for longer periods of time and cover a wider variety of relevant issues than was the case with the Pittsburgh group.

The same five cognitive structure probes were used in pre- and post-instruction. Thus far, the analyses of these data have been somewhat different from the Pittsburgh analyses, being more global, having a stronger focus on the individual probes than on the group probes, and including inspection of the transcripts of the instruction sessions. The major reasons for this approach being the first consideration of the data are the very different pre-instruction cognitive structures of the group (as compared with the Pittsburgh group) and the clear indications through the period of the study of changes in the ways individuals interpreted physics-based events and issues (and, therefore, the implication that cognitive structures had been changed).

FINDINGS AND DISCUSSION

All six students had undertaken physics at Years 11 and 12 in high school (the normal mode of school physics study in Victoria), and four had completed first year university physics. All were biology or chemistry majors. Four claimed to understand basic mechanics when given open-ended questions about their knowledge of areas of physics at the first sessions, one stated he did not understand vectors and another that he did not understand mechanics but did understand forces. Given the extent of physics courses undertaken, it would be expected that the students would have some knowledge of school physics. Within limitations this was the case. Where individuals are identified in the discussion, C, I, J. L, N, and Z are used.

Data from the word association task did not show the characteristics evident in the Pittsburgh data. At the analysis-by-inspection level used thus far, much more consistency is evident. Just over 50% of pre-test associations were repeated in the same sense (as judged by the sentence produced for each response) on the post-test. A significant number of non-repeated responses were synonyms and examples (e.g., in response to *change of motion—stop, start, slow down, veer;* to *force—gravitational, electric, magnetic, nuclear*). Although the pre-test produced a considerable number of propositions, there were relatively few relationships. Associations reflecting $F = ma$ were given by all, but the accompanying propositions indicated a lack of precision in the view of this relationship. The above comments do not apply to J whose associations were more precise and reflected something nearer to a physicist's view.

In every case but one (student I), the first chosen pair of concepts in the tree construction task were descriptive and kinematic (e.g., speed–velocity, displacement–distance). Student I began with force-resultant. This observation bears striking similarity to the pairs selected by the younger Pittsburgh sample, which were overwhelmingly descriptive and kinematic. This observation is interesting in terms of the view expressed by Easley (1971; quoted at the beginning of the paper). In the free sort task there was a tendency for groupings to be formula-based.

The request for definitions which is part of the ConSAT Task caused difficulty for all students. While all students sorted all concepts into the "recognized" pile, except for N who had two "unrecognized" (frame of reference, inertia), there was universal unease about the request for definitions. Although demonstrably incorrect definitions were rare, attempts which were sufficiently vague as to be inadequate were relatively common. Even vagueness was not always reached for more abstract concepts such as inertia. The subsequent concept maps, in most cases, reflected these problems. Inertia was most commonly placed on the map with considerable difficulty. Links beyond the initial joining of two concepts were rare. As for the word association task, student J was an exception. Her concept map was more integrated and her definitions both more precise and more confidently advanced.

All students gave predictions consistent with subsequent observations for situations in the DOE interview. However only one (student J) consistently used a force analysis in explaining this series of statics and dynamics situations. The other five, to varying degrees, explained in terms of the surface features of the demonstration rather than the underlying physics. Both these explanations and the reactions to the set of provided answers produced a number of incorrect statements about forces and motion.

In general, with the exception of J, the pre-instruction cognitive structures of the students were characterized by the presence of relationships in the form of formulas which were not utilized for real events or other applications and by the presence of considerable propositional knowledge which lacked specificity and was occasionally incorrect. A very brief summary of these features emerged at the beginning of the first instructional session. The first task in this session was to consider the question, "What are some of the things we can say about the relationships between force and motion?" Students recorded consensus answers on a blackboard. A facsimile of that blackboard is given in Figure 11.1. This blackboard summary remained throughout the period of instruction, with modifications being made as students saw these to be necessary. It is interesting to note that an early

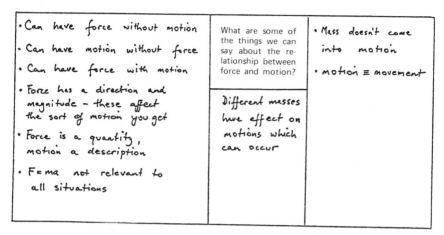

FIGURE 11.1   Blackboard summary after first segment of first instructional segment.

response given to the beginning question in the first session was "$F = ma$."
All agreed this was an appropriate answer but, when then asked to explain
what it meant, no student was prepared to volunteer an answer.

In their reactions to what was happening written after each instructional
session, all students made claims which suggest that changes were occurring
in their cognitive structures. Their reactions are summarized before the
post-instruction probes are discussed.

After the first session, all wrote in terms of "I don't understand what I
thought I did," and all indicated a positive motivation to the experience.
This positive reaction was apparently contributed to by a long discussion
about normal reaction forces. During this discussion (as shown by the transcript), considerable steps were taken toward reconciling an intuitive view of
force being a dynamic entity associated with movement with existing knowledge from physics courses that a normal reaction force should be considered.
In other words, as well as having their beliefs about their own understandings shaken, they also perceived that their understanding had increased in
one area. The second session was largely devoted to prediction, observation,
and discussion of falling bodies. (The actual instruction in this session is
discussed more fully in the conclusion.) Student reactions after this session
indicated considerable change in their views about physics learning:

[The session] enabled me to see how others view things and why they view them this
way. Made one think hard to get a totally convincing argument for your side and any
inability to do this gives you the suspicion that you are not in fact correct in your
initial explanations . . . helps in understanding our false misconceptions and being
confident in our new ideas. (student N)

During [the] group discussion I changed my mind about the forces operating on the falling bodies. . . . Some people fight hard not to change preconceived ideas. (student I)

It feels strange to contradict oneself half an hour later. It's worth the lengthy time involved because I can have time to gradually understand the issue at point. . . . [It's] as if we are trying to turn a blind eye to the truth. It's comforting to try to keep certain ideas forever even if there's a chance that they may be wrong. (student Z)

I'm shattered! Didn't realize how devastating it could be to have a deep rooted belief proved wrong. Can I blame my physics teacher? It would be all right if some dummy didn't pose a question which could be used to support the opposite argument. Seriously though, very instructive. I don't know if I'm going to be able to last the distance. I'm mentally exhausted after each session and the effort to hold out when I'm wrong is very draining. Great fun so far even if I hate it at odd times. (student C)

Even though I had the correct idea today that acceleration is approximately constant, at first I could only say that the forces were different by using $F = ma$. After having to think of arguments for this, I could justify it from an observational point of view. . . . The open discussion and justification of ideas is a brilliant way toward understanding physics. Substituting numbers in formulas seems to be a poor alternative. (student L)

Student J, whose pre-instruction data suggested a more comprehensive relevant cognitive structure, gave physically appropriate predictions and explanations for all situations used in Session 2. However, even she reported that "my *very first* impulse was to claim that equal forces were acting on the balls [of different mass]."

The general trend evident in the above quotes continued through the remaining session, although the reactions were not elaborated in such detail. For example, student C commented, "I don't smart so much any more" (after Session 4) and, "Becoming more receptive to change in my supposed convictions" (after Session 5). The written reactions to the whole sequence of instruction were also very consistent with the above quotes from Session 2 reactions. Students all expressed the belief that their views had changed (with all giving detail of at least some of these changes). The importance of discussion, of considering the views of others, or relating a situation under consideration to other real-world phenomena were all seen to be significant in promoting changes of view.

The blackboard summary at the end of the instruction is represented in Figure 11.2. The progressive changes evident in this figure had been made in response to the question, "Is there anything you would like to change on the blackboard?" at the end of Sessions 2, 4, and 5. Students were not specifically asked to add additional statements, only to consider modifying what was there. This final summary shows less imprecision and more correctness (in terms of the physicist's view of force and motion).

The post-instruction probes of cognitive structure suggest that the cog-

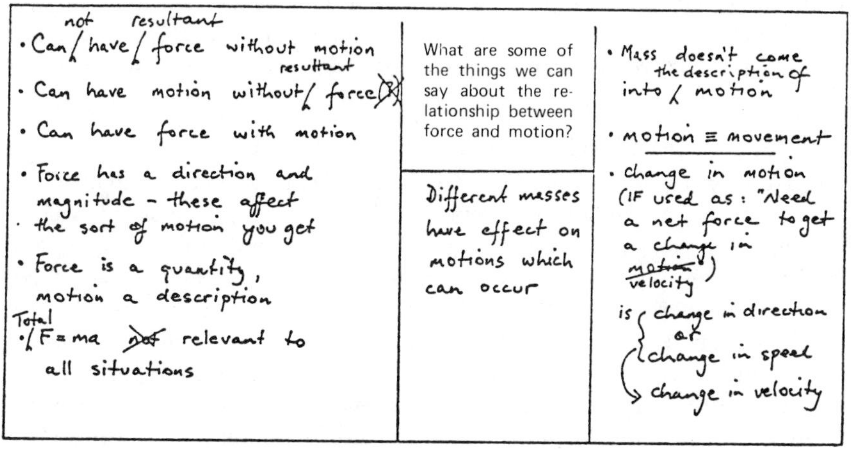

• Can not have resultant force without motion
• Can have motion without resultant force
• Can have force with motion
• Force has a direction and magnitude - these affect the sort of motion you get
• Force is a quantity, motion a description
• Total F = ma relevant to all situations
What are some of the things we can say about the relationship between force and motion?
Different masses have effect on motions which can occur
• Mass doesn't come into the description of motion
• motion ≡ movement
• change in motion (IF used as: "Need a net force to get a change in velocity")
is change in direction or change in speed
change in velocity

FIGURE 11.2 Blackboard summary after completion of instruction.

nitive structures of four of the students have been changed in the ways they reported through the instruction. The exceptions are student J (who, in simple terms, had less to change in her cognitive structure), and student N.

The most powerful, even though general, evidence of a change in cognitive structure comes from the DOE interview. All students except N immediately described the situation used as being the same as that used in the pre-test despite the two situations having quite different surface features. All but N consistently used the underlying physics principles to explain their predictions and their evaluations of the provided set of answers. Student N showed some tendency to do this, but was not consistent. Only student N used imprecise statements in responding to the situations.

The definition task involved in the ConSAT Task was much more positively approached and much more appropriately answered. Impression was evident in mass and inertia definitions from student N and inertia from student C. By comparison with the pre-test maps, concept maps showed more connections and greater precision in connections.

In the tree construction task, the most related pair of concepts was seen to be *mass* and *inertia* by students I, J, and L. The remaining three selected a pair of descriptive, kinematics terms as they had done on the pre-test, although no student began with the same pair as they had on the pre-test. The word association data showed a decrease, relative to the pre-test, in synonym and example responses for all students. However, the total associations given increased by 29%, and by about 40% for the abstract concepts *force*, *mass*, and *inertia*. Increases were much smaller for the more familiar

178   AUDREY B. CHAMPAGNE, RICHARD F. GUNSTONE, AND LEOPOLD E. KLOPFER

descriptive terms speed and change of motion. Incorrect propositions on this task were very infrequent and there was much more precision evident. Relationships between concepts were more frequent than on the pre-test.

FURTHER PITTSBURGH ANALYSES

In addition to examination by inspection of data collected from the instructed group, a number of statistical analyses which apply various scaling techniques to the response data obtained via the several probes of cognitive structure have been undertaken.[1] The data from the three group-administered probes (word association, free sort, and tree construction tasks) have been analyzed. Proximity matrices have been produced from responses to each of the three tests and scaling methods applied to these matrices to produce representatives of cognitive structure. Latent partition analysis (LPA), multi-dimensional scaling (MDS), and hierarchical clustering (HC) methods were applied to the proximity matrices from each of the three tests, with the exception of the word association matrix to which LPA cannot be applied. Hence eight representations of group cognitive structure resulted. The procedures adopted for analysis of responses to each of the three tasks is described elsewhere (Champagne et al., 1984). A brief discussion of the results of the quantitative analysis follows.

Analysis of the Pittsburgh data from the word association task yielded relatedness coefficients for pre- and post-test data from the same group that are generally low and sometimes zero. This finding led us to conclude that these input data were too unreliable to permit valid statistical analyses. Hence no further analyses are reported.

The previous finding is difficult to interpret. Both in a previous study in which an identical word association task was used in a pre- and post-instruction context (Gunstone, 1980) and in the Victorian section of this study quite contrary observations have been made. Because these latter uses involved Grade 11 and teacher-training students respectively, the age of the Pittsburgh students might provide an explanation. It is logically reasonable to suggest that the Pittsburgh word association observations result from students of this age having cognitive structures in this content area which are, relative to older students, poorly articulated and unstable in terms of propositional knowledge. However, no support for this hypothesis exists beyond the general perspectives we have formed from working in this content area with middle-school students.

The other analyses conducted suggest little change in the cognitive structure representations derived for the Pittsburgh students. For every com-

[1]These analyses were developed primarily by our colleague, Ron Hoz, who also guided the data reduction, scaling, and interpretation work.

bination of the two remaining data-gathering tasks (free sort, tree construction) and three scaling methods (LPA, HC, MDS) used: (1) the cognitive structure representations of the instructed and control groups derived from the pre-test are quite similar, and (2) the cognitive structure representations of the control group derived from the pre- and post-tests are quite similar. The degree of similarity in these two comparisons serves as a baseline for assessing possible pre-post- changes between pre- and post-test in the cognitive structure representations of the instructed group students. We find that the instructed group's post-test representations are just as similar to their pre-test representations as the representations compared in comparisons (1) and (2) are. Hence, changes are not evident in the instructed group students' cognitive structure representations from pre- to post-test. Using the same baseline, we also observe that, consistent with the preceding findings, the cognitive structure representations of the instructed and control groups derived from the post-test are quite similar. Only one set of representations is presented here, as this set is typical of the consistency evident in all representations. Figures 11.3–11.6 show two-dimensional con-

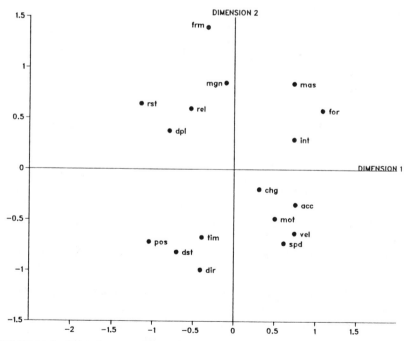

FIGURE 11.3   Rotated 2-D configuration of free sort task, instructed group, pre. (Abbreviations used: acc, acceleration; chg, change of motion; dir, direction; dpl, displacement; dst, distance; for, force; frm, frame of reference; int, inertia; mas, mass; mgn, magnitude; mot, motion; pos, position; rel, relative; rst, resultant; spd, speed; tim, time; vel, velocity.)

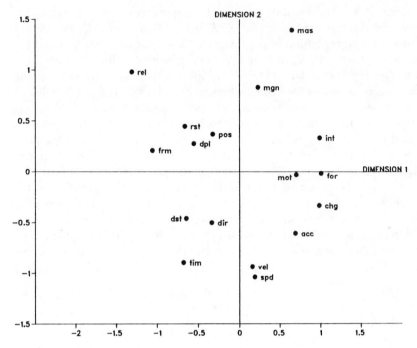

FIGURE 11.4   Rotated 2-D configuration of free sort task, instructed group, Post. (Abbreviations explained on Figure 11.3.)

figurations derived from the free sort task data for instructed group pre- and post-test and control group pre- and post-test, respectively. These representations depict the 17 concepts located in a two-dimensional space, which was chosen as the optimal solution in the application of multidimensional scaling. Both visual comparisons and calculated indices of goodness of fit for each pair of representations suggest considerable stability in the cognitive structure representations of the instructed group.

This apparent failure of the instruction to produce detectable changes in cognitive structure has a number of possible interpretations, three of which we advance here.

1.   Our previous work points to the difficulty of changing cognitive structures associated with the interpretations of events which are part of the students' real world. In the area of introductory mechanics considered in this study in particular, there are grounds for arguing that pre-instruction conceptions derived from experience are more resilient than are conceptions in other science areas.

2.   The possible explanations for the unexpected characteristics of the Word Association Task data presented above may be relevant here. That is,

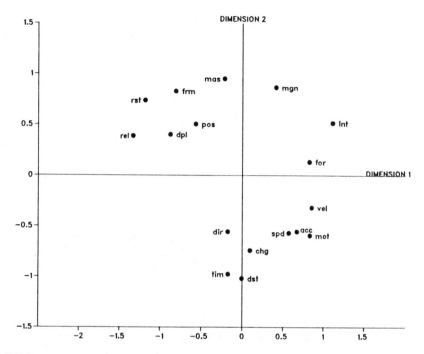

FIGURE 11.5   Rotated 2-D configuration of free sort task, control group, Pre. (Abbreviations explained in Figure 11.3.)

the failure to detect change in these cognitive structure representations may be a result of students of this age having cognitive structures in this content area which are not well articulated and are frequently unstable in terms of propositional knowledge. Given this, one could argue that the stability in representations may result from the nature of the task, and might not be found in other data.

3.   Although no change was detected, there were differences in the instructed group pre- and post-test responses to the cognitive structure probes. For example in the Free Sort Task, differences are observable both by inspection and by the correlations of the proximity matrices derived from the pre- and post-test responses of both groups. The correlations may indicate changes in the instructed group cognitive structure that were not detected by the target terms and analytical methods applied.

The target terms were selected to represent the major concepts related to classical mechanics and to detect any major restructuring of the students' conception of mechanics from some naive perspective to the Galileo-Newtonian one. A significant finding of the quantitative analysis is the similarity of cognitive structures of the instructed and control groups, both before and

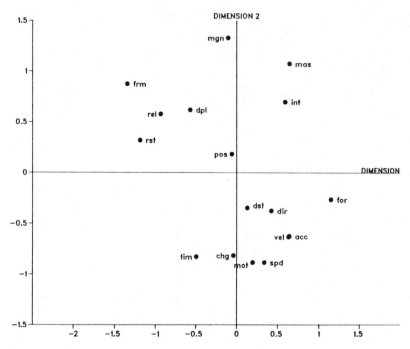

FIGURE 11.6   Rotated 2-D configuration of free sort task, control group, Post. (Abbreviations explained in Figure 11.3.)

after the instructional treatment of the instructed group. This finding lends support to the qualitative studies done by many investigators which have revealed the consistently Aristotelian characteristics of the naive conceptions of mechanics of a broad range of students. These results also substantiate the reports by other investigators that the Aristotelian features of beginning students' conceptions are difficult to change.

Clearly no major restructuring of the Pittsburgh students' conceptions occurred as the result of the instruction. However, there is evidence that students' knowledge about certain target terms (e.g., frame of reference) did increase and that this knowledge made the students' analysis of the motion of objects more powerful. Some important changes also occurred in students' differentiation of terms; for example, their awareness that there is a significant difference between mass and weight. These comments about changes in instructed students' knowledge are consistent with findings in the quantitative analysis of the data (Champagne et al., 1984). Moderate differences between proximity matrices derived from task response data are washed out by the analytical methods used to produce the cognitive structure representations.

CONCLUSION

The purpose of this research was to investigate the effectiveness of the ideational confrontation strategy in producing conceptual change. Conceptual change can be effected in either the contents of cognitive structure or their structural organization. Findings from this preliminary study suggest that the ideational confrontation strategy has promise both for engendering restructuring of students' existing knowledge and for changing the contents of their knowledge. Moreover, the strategy produces other significant changes, for which we saw evidence in the complementary findings for the two groups.

A general observation from the findings is that the effects of the ideational confrontation strategy differ somewhat for students with different characteristics. Perhaps this interaction was most clearly demonstrated with respect to the changes in knowledge structure and contents, but it applies to the other aspects of change as well. An obvious difference between the students in our two groups is that the Victorian students had more verbal knowledge derived from formal physics at the outset than the Pittsburgh students. The contents of the Victorian students' knowledge before the instruction included quite a reasonable share of propositions based on the Galileo–Newtonian view of mechanics (even though their interpretations of real phenomena were often not Newtonian). By contrast, the knowledge about motion held by the Pittsburgh students, who as a group had previously had very little mechanics-related instruction, was essentially limited to their naive, Aristotle-like, experience-based conceptions pertaining to the motion of objects. In the course of instruction utilizing the ideational confrontation strategy, the Victorian students were able both to place a new structure and to better reconcile this formal knowledge with interpretations of real-world phenomena. For the Pittsburgh students, the main effects of the strategy on their knowledge were to increase its precision through the differentiation of related terms, to augment the contents with some new mechanics concepts (e.g., resultant) and to inject the physicist's canonical meanings of certain concepts (e.g., acceleration, force). All these effects can contribute to the formulation of explicit propositions about motion and to the nascent structuring of propositions, but they do not indicate a restructuring of the student's larger conceptual structures related to the motion of objects. In other words, the effects of ideational confrontation on the Pittsburgh students' knowledge differs from those for the Victorian students, but they are also consistent with the differences in initial knowledge.

Another general area where the strategy produced complementary effects for the two groups concerns the students' construction of valid and convincing explanations for observable physical phenomena. It is remarkable that the students in both groups were initially quite reluctant to construct such explanations, as we saw in the pre-instructional DOE interviews and the initial days' discussions. For the Victorian group, the change from reluctance about explaining motion phenomena to persistence in seeking the best explanation took place more rapidly than for the Pittsburgh group and was closely related to the developing awareness of their own thought processes and the nature of their knowledge. Inspection of the transcripts of the Victorian instruction sessions and the students' reactions to the sessions does not suggest a dramatic "flash of inspiration" when this change was suddenly made, but rather that the students' self-awareness developed over a period of time. However, it is clear that the ideational confrontation in the second session was of great significance. In this session, students began by predicting the relative times of fall for 2 spheres of about 0.1 m diameter when dropped from about 2 meters. One sphere was lead, the other a plastic ball. Although all predicted equal times, four (C, I. N, Z) said this would be the case because the same force acted on each sphere. This difference was debated at length. The process was then repeated for the spheres dropped from a height of about 12 meters (for which there is unquestioned difference in the time of fall), for two cubes of identical volume and different mass dropped about 0.25 meter in water, and finally for the two cubes dropped the same distance in oil. This process occupied about $5\frac{1}{2}$ hours (with coffee and lunch breaks not included). Transcripts and students' reactions make it clear that the first part of the next session—discussion of a demonstration of two carts of different mass being pulled by identical forces at the same time—should be seen as part of this sequence.

The significance of this sequence lies not in the changes in specific knowledge contents per se, but in the way in which this experience promoted strongly held alternative views and lengthy and voluble debate. This particular sequence resulted in all students indicating that they began to understand the origin and resilience of views which caused them to be conceptually troubled by the perspectives advanced in their previous physics courses. It was common through the remainder of the instruction for individuals to overtly use the second session experience to argue to themselves and others about internal consistencies in their interpretations of events and inconsistencies between these interpretations and the tenets of Galileo–Newtonian mechanics. Having developed this degree of introspection, the Victorian students became quite adept in seeking out reasonable explanations for motion phenomena and, in doing so, they were prepared to continue debate on an issue for extreme lengths of time.

The Victorian students' far greater maturity and cognitive development allowed them to become more introspective than the younger Pittsburgh students. Nevertheless, at their own level of development, the Pittsburgh students also learned to devise explanations for motion phenomena through the process of constructing arguments. In fact, the ideational confrontation strategy requires that students become skillful in developing arguments to support their assertions and to defend a point of view in the face of conflicting opinions. Students must also be motivated to exercise these skills. A major change in the behavior of the students in the Pittsburgh group was the development of the ability and the willingness to engage in discussions about the motion of objects.

Once the students became convinced that this behavior was acceptable— even desirable—the group debated scientific issues at length and with enthusiasm. Our preliminary analyses of transcripts of the instruction suggest that the quality of the students' arguments improved over the course of the instruction. At about the mid-point of the Pittsburgh instruction, a visitor observed a lengthy discussion about the weight of a brick in water, and he expressed both surprise and delight about the length and quality of the discussion. Comments made by two of the students' teachers who also observed the discussion are also significant. One teacher noted that much valuable time would have been saved if the students had simply been told the answer. Another teacher wondered if the science classroom is an appropriate place for discussion. Given the prevailing attitude of teachers, the observed initial reluctance of the students to engage in discussion that lead to reasoned explanations is not surprising. However, the important observation is that this reluctance can be overcome via the use of such strategies as ideational confrontation.

One problem noted with regard to this strategy was the large difference in participation in the discussions by the Pittsburgh young women and men. With one exception, there was significantly less participation in the discussion by the young women, and efforts by the female instructor to actively encourage their participation were not successful. At this time, we have no satisfactory suggestions for alleviating this problem, which probably has its roots in prevailing societal attitudes and in the developmental process of adolescents. The problem of differential participation in discussions by gender was not strongly manifested in the Victorian group of much older university students, all of whom were committed to becoming teachers, whose social role is generally perceived as expressive.

In summary, we have successfully used the ideational confrontation strategy to produce conceptual change in a content area for which we have found this process to be very difficult. We have achieved this change in what we regard to be the remarkably short time of approximately 30 instructional hours, spread over 5 days (Victoria) or 8 days (Pittsburgh). The changes in both groups involved the development of some understanding of the scientist's perspective concerning the physical domain of motion, an understanding which may well be applied by the students in other content areas.

Especially the Victorian students, with their greater introspective skills, were able to gain considerable insight into the distinction between the scientific way of knowing and knowledge based on individual experience and interpretation, and to use this insight to gain considerable understanding of the processes of their own learning. On completion of the course, all the Victorian students claimed that these personal metacognitive insights would influence their subsequent learning. This claim has not been investigated.

This research also has implications for the psychological theory underlying the ideational confrontation strategy. Ideational confrontation is one version of interactive dialogue, which has been proposed as a strategy to facilitate schema change (Anderson, 1977; Riegel, 1973). However, advocates of the strategy have not posited any theoretical mechanism by which the strategy produces cognitive changes. The results of the study reported here may inform the process of the development of such a theory. We propose that the active mental manipulation of concepts and principles in the formulation and revision of arguments plays an important role in the differentiation and integration of concepts and in the restructuring of knowledge. Equally significant is the strategy's influence on the development of the distinction between the scientific way of knowing and knowledge based on individual experience and interpretation. The process of engaging in interactive dialogue helps develop the ideas of rules of evidence and criteria for validity. This knowledge about the nature of science concepts and the procedures involved in establishing them may be a crucial link in the learner's formulation and reformulation of the contents of his or her own knowledge. This kind of mediating link may be developed in the course of engaging in interactive dialogue, such as ideational confrontation.

REFERENCES

Anderson, R. C. (1977). The notion of schemata and the educational enterprise: General discussion of the conference. In R. C. Anderson, R. J. Spiro, & W. E. Montague (Eds.), Schooling and the acquisition of knowledge. Hillsdale, NJ: Erlbaum.
Champagne, A. B., Hoz, R., & Klopfer, L. E. (1984, April). Construct validation of the cognitive structure of physics concepts. Paper presented at the meeting of the American Educational Research Association, New Orleans.
Champagne, A. B., Klopfer, L. E., & Anderson, J. H. (1980). Factors influencing the learning of classical mechanics. American Journal of Physics, 48, 1074–1079.
Champagne, A. B., Klopfer, L. E., DeSena, A. T., & Squires, D. A. (1981). Structural representations of students' knowledge before and after science instruction. Journal of Research in Science Teaching, 18, 97–111.
Champagne, A. B., Klopfer, L. E., & Gunstone, R. F. (1982). Cognitive research and the design of science instruction. Educational Psychologist, 17, 31–53.
Champagne, A. B., Klopfer, L. E., Solomon, C. A., & Cahn, A. D. (1980). Interactions of students' knowledge with their comprehension and design of science experiments (1980/9).

Pittsburgh, PA: University of Pittsburgh, Learning Research and Development Center Publication Series. (ERIC Document Reproduction Service No. ED 188 950)

Deese, J. (1965). *The structure of associations in language and thought*. Baltimore: Johns Hopkins.

Easley, J. A. (1971). Scientific method as an educational objective. In L. C. Deighton (Ed.), *The encyclopedia of education*. New York: Macmillan & Free Press, 8, 150–157.

Fillenbaum, S., & Rapoport, A. (1971). *Structures in the subjective lexicon*. New York: Academic Press.

Gorodetsky, M., & Hoz, R. (1985). Changes in the group cognitive structure of some chemical equilibrium concepts following a university course in general chemistry. *Science Education, 69*, 143–152.

Gunstone, R. F. (1980). *Structural outcomes of physics instruction*. Unpublished doctoral dissertation, Monash University, Melbourne, Australia.

Gunstone, R. F. (1981, April). *Cognitive structure and performance after physics instruction*. Paper given at the meeting of the American Educational Research Association, Los Angeles. (ED 202 703).

Gunstone, R. F., Champagne, A. B., & Klopfer, L. E. (1981). Instruction for understanding: A case study. *Australian Science Teachers Journal, 27*(3), 27–32.

Gunstone, R. F., & White, R. T. (1981). Understanding of gravity. *Science Education, 65*, 291–300.

Johnson, P. E. (1965). Word relatedness and problem solving in high school physics. *Journal of Educational Psychology, 56*, 217–224.

Johnson, S. C. (1967). Hierarchical clustering schemes. *Psychometrika, 32*, 241–254.

Miller, G. A. (1969). The organization of lexical memory: Are word associations sufficient? In G. A. Talland & N. C. Waugh (Eds.), *The pathology of memory*. New York: Academic Press, 223–236.

Preece, P. F. W. (1976). Mapping cognitive structure: A comparison of methods. *Journal of Educational Psychology, 68*, 1–8.

Rapoport, A. (1967). A comparison of two tree-construction methods for obtaining proximity measures among words. *Journal of Verbal Learning and Verbal Behavior, 6*, 884–890.

Riegel, K. F. (1973). Dialectic operations: The final period of cognitive development. *Human Development, 16*, 346–370.

Shavelson, R. J. (1972). Some aspects of the correspondence between content structure and cognitive structure in physics instruction. *Journal of Educational Psychology, 63*, 225–234.

Shavelson, R. J., & Stanton, G. C. (1975). Construct validation: Methodology and application to three measures of cognitive structure in physics instruction. *Journal of Educational Measurement, 12*, 67–85.

Wiley, D. E. (1967). Latent partition analysis. *Psychometrika, 32*, 183–193.

METALEARNING AND METAKNOWLEDGE STRATEGIES TO HELP STUDENTS LEARN HOW TO LEARN*

Joseph D. Novak

INTRODUCTION

For centuries, all serious teachers have recognized that students who learn rapidly and who can retain and use their knowledge in new contexts are not common. The question has been, why do some students learn so well and others so poorly? This challenge was met by Socrates with the strategy of "Socratic questioning" wherein through a clever sequence of carefully chosen questions, even the ignorant slave could be brought to understand the world as Socrates saw it. The assumption was that knowledge existed covertly in all humans and that proper questioning could reveal this knowledge in a way analogous to the way in which human potential was revealed when the form provided by the father was freed to develop by copying itself onto the material provided in the mother's womb. No one believes today that the form of the human expands from the miniature folded in the sperm and very few people believe that the knowledge exists preformed in the human brain and the secret to intelligence is to find strategies to let it out. However, I argue that there is great learning potential in humans that remains undeveloped and that many common educational practices impede rather than enhance expression of this potential. Current knowledge about human learning (metalearning) and knowledge about the processes by which humans construct new knowledge (metaknowledge) can help to release much more of the intellectual potential of humans. Some recent research completed at Cornell University is cited to support this thesis.

*An earlier version of this chapter was presented at the International Seminar on Misconceptions in Science and Mathematics, Cornell University, June 21, 1983.

PSYCHOLOGICAL BASIS FOR METALEARNING
AND METAKNOWLEDGE

Since 1964, our research program has been based upon the cognitive learning theory of David Ausubel. His *Psychology of Meaningful Verbal Learning* (1963) and later *Educational Psychology: A Cognitive View* (1968) present a comprehensive, coherent theory of cognitive learning that is explicitly directed toward *human* learning, especially in school settings. Over a decade or so of application in research and instructional planning, we gradually came to understand his theory, and this work also led to some modification of the theory (Ausubel, Novak, & Hanesian, 1978). Published in six languages, Ausubel's theory has had world-wide recognition but probably has the least acceptance in the United States. Nevertheless, our research group continues to find his theory, with a primary emphasis on the nature of *meaningful* learning, the most powerful and comprehensive for our work. We have added some aspects of "cognitive science" to the theory and further modified the theory in regard to cognitive development (Novak, 1977a, 1977b; 1980; 1982). Wittrock's (1974) Generative Learning theory had its origins in Ausubelian cognitive psychology and Wittrock has elaborated on some of Ausubel's earlier ideas. Mayer (1983) has summarized some of his early work with Ausubelian theory, but in his *Promise of Cognitive Psychology* (1982), he chose to ignore the early contributions in deference to now widely popular "cognitive science" views.

With regard to cognitive development, my current view is that by age 3, all normal children have essentially the same cognitive *operational* capacity as adults, but what varies from person to person and with increasing years are the frameworks of *disciplinary specific* concepts and propositions individuals possess. Our position is supported by recent work as that of Kiel (1979) and Macnamara (1982), plus our own interpretation of a large body of research dealing with human cognitive performance (see Novak, 1977b). However, I argue in this Chapter that generically relevant cognitive learning strategies can be acquired and that it is possible that important *qualitative* differences develop in learners over the span of school years. In fact, we have suggested that these qualitative differences may account for much of the gross underrepresentation of women in science and mathematics careers (Ridley & Novak, 1983).

The key idea in Ausubel's theory is the nature of meaningful learning, as contrasted with rote learning. Ausubel's distinction here is at once simple and profound. He defines meaningful learning as non-arbitrary, non-verbatim, substantive incorporation of new knowledge into a person's cognitive structure, whereas rote learning is described as arbitrary, verbatim, non-substantive incorporation of new knowledge into cognitive structure.

Taken by themselves, each of the words in the previous sentence is potentially obvious. We all know that verbatim learning occurs when we memorize definitions (e.g., a noun is the name of a person, place, or thing) without stopping to consider the *meaning* of each word in the definition and their combined meaning. We all have experienced "losing points" on a teacher's test when our definition was not *verbatim* the teacher's definition. We may have provided a *substantively* identical definition, but we chose different words or word sequences to do it. We may have regarded the teacher's wording as *arbitrary*, but the teacher regarded it as sacrosanct. In religious or mystical rituals, verbatim repetition may be sacrosant, but this should rarely be essential in school learning. Ausubel goes further to define meaningful learning as requiring (1) meaningful learning materials (Ebbinghaus did studies on retention following rote learning of nonsense material), (2) a meaningful learning set, for example, a disposition on the part of the learner to link each concept label in the new material with concepts he or she already possesses, and (3) *relevant* cognitive structure, that is, some concepts already present in cognitive structure that can be related non-arbitrarily to the new concept labels.[1] As Macnamara (1982) has shown, young children do this remarkably well between the ages of 15 and 30 months. They also do something else that is remarkable—all normal youngsters *discover* autonomously the meanings of some abstract concept labels such as *I* or *you*, or *she* or *me*. Subsequently, they use these discovered concept labels to acquire from older children or adults the meanings of new concept labels, and vocabulary building accelerates rapidly until early school years. Macnamara found that his son, Kirnan, acquired meanings for 297 words (concept labels) in his fifteenth through nineteenth month but meanings for 211 words in his twentieth month.

In case there are still any Skinnerians around, I would point out that to use 211 words without errors in simple two word sentences, Kirnan would have had to form $\frac{211!}{2!}$ stimulus–response linkages during his twentieth month. Of course, not all two word sentences make sense, so we might divide this number by, say, a million—a very generous assumption that only one in a million sentences with only two words Kirnan knows make English sense. I have not found a computer that will compute the number for 211 factorial, but it is essentially infinity. Even divided by 30 million, Kirnan would still have had to learn a very large number of new English phrases each day! So we return to the more plausible theory that the neurological hard wiring of humans permits them to acquire meanings for concepts and to relate these meanings in almost infinite varieties of ways. We do not yet know the neurobiology of this process, but then Mendel did not know that base pair

[1]We define *concept* as a perceived regularity in events or objects designated by a label. The remarkable capability "hard wired" into all normal human nervous systems is the facility to perceive regularities and to use language labels to encode these regularities. This is as much an evolutionary achievement of *Homo sapiens* as bipedalism. It does not have to be learned.

sequences in DNA were determining the color, shape, and size of his pea plants. We can apply valid psychological principles of learning without knowing the biochemical bases of these principles.

The principle of meaningful learning includes the idea that each of us has a unique sequence of learning experiences and hence each of us acquires idiosyncratic meanings for concepts. For this reason, Ausubel chose to use the word *subsumer* or subsuming concept to designate the functional unit in memory of each person. Each culture has more or less common meanings for the word labels for concepts, but each individual's subsumers are in at least small ways idiosyncratic. In some instances, this idiosyncratic meaning departs widely from the culturally accepted meaning and we say the person has a misconception or alternate framework. Once established in cognitive structure, these idiosyncratic subsumers are not easily modified, as is well documented by recent studies (see Helm & Novak, 1983).

Students of foreign languages are familiar with the fact that word for word translations to or from English are sometimes difficult, even when the foreign word designates more or less the same regularity in events or objects as an English equivalent. The epistemology of concept labels tells us they are tied not only to the events or objects referred to by the label but also to the whole context in which those events or objects are experienced. Often both the social and the physical context are so different that good translations are not possible. The classic case here would be from the English *snow* to Eskimo, where several words (concept labels) exist for snow. Hewson (Chapter 10, this volume) expands on the idea that our culture can be a substantive influence on the way we perceive regularities in the environment and their relationship to the language labels we use for concepts.

Three other basic ideas form the core of Ausubelian learning theory and together serve to explain most cognitive phenomena. As new knowledge is acquired through meaningful learning, subsuming concepts undergo *progressive differentiation;* that is, new events or objects or concepts labeling regularities in new events or objects are seen as substantively related in the form of new propositions[2] that include the original subsumer. Working from Ausubel's theory and current ideas from epistemology, we have developed a strategy we call "concept mapping" that can be used to illustrate progressive differentiation of concept meanings by the addition of new propositions (see Figure 12.1).

[2]Propositions are two or more concepts linked in a semantic unit, for example, sky is blue, sky is air, sky is molecular. Propositions are the "molecules" from which meaning is built and concepts are the "atoms" of meaning, to use a rough metaphor. The English language contains about 450,000 concept labels, and these can be combined into an infinite number of meaningful propositions.

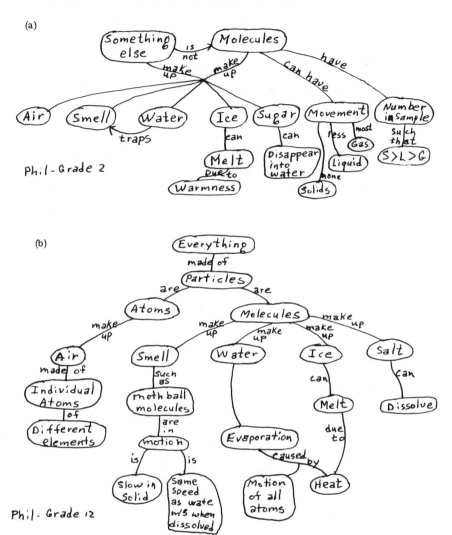

FIGURE 12.1   Two concept maps drawn from interviews with a student in grade 2 (a) and grade 12 (b). Note that confusion persists regarding the atomic-molecular composition of matter even after junior high school science and high school biology, chemistry, and physics.

Another key idea is that occasionally the meanings of two or more concepts will be seen as related in a new and significant way and *integrative reconciliation* takes place. For example, a student may learn that the concept of *space* is related to the structure of solids, liquids, and gases and then recognize that it is the amount of space between molecules that accounts for the expansion of substances, and not a "fattening" of the component molecules. The concepts of *space* and *molecule* now take on new meaning. Whenever integrative reconciliation occurs, all of the concepts in the reconciled propositional framework take on at least slightly altered meanings. Occasionally, creative persons construct some novel integrative reconciliation (such as Einstein's $E = mc^2$) and our whole view of a segment of the world changes. I equate creativity with the ability and emotional proclivity to form high order integrative reconciliations. Rote learning not only fails to provide the basis for the latter but tends to inhibit search for integrative reconciliation—as does much of school learning and evaluation. Human creative potential is probably at least an order of magnitude greater than what is now manifest, as a result in part of the inhibiting effect of most school practices. Unfortunately, women are especially prone to be socialized into playing the school game and this may account in part for their underrepresentation in the ranks of creative giants (Ridley & Novak, 1983).

Occasionally, a new concept meaning is acquired that subsequently serves to integrate the meaning of two or more concepts. Such superordinate learning is not common partly because disciplines do not have large numbers of truly superordinate concepts and partly because subsumption appears to be an easier learning pathway. When superordinate learning does occur, one or several instances of integrative reconciliation may soon follow. For example, the chemistry student who acquires the meaning of entropy sees solubility, reaction rates, and osmosis with substantively new meanings. To see what such superordination can do in a discipline, think about what relativity has done for physics or plate tectonics for geology.

There is more to Ausubel's theory, and, of course, as a living theory, it continues to be modified by new research. For example, our studies led to the necessity of including progressive differentiation and integrative reconciliation as key learning principles, whereas Ausubel's earlier work (1963; 1968) presented these ideas as principles of instructional design. A principle belonging to the latter category is Ausubel's advance organizer, but much writing erroneously tends to at least partially equate this principle with his theory of cognitive learning.

PHILOSOPHICAL BASIS FOR METALEARNING AND METAKNOWLEDGE

In my view, one of the tragedies influencing education has been the adherence of psychological and educational research to an outdated philosophical foundation. When Bacon proposed in 1620 that scientists eschew preconceived notions and ideas and devote themselves to unbiased, objective observation, he was properly counseling against use of the mystical ideas that were held by many natural philosophers ("scientists") of his day. Bacon laid the foundations for empiricism or the notion that carefully observed and recorded reality can lead to true knowledge about how the world works. Psychologists, especially behaviorists, have rigidly adhered to this view for the past century. Most educational research that has grown out of this tradition also attempts rigid empiricism, albeit, this is much harder to do in classrooms than in the laboratory. Unfortunately, as Toulmin (1972) and more recently Popper (1982) have tried to point out, empiricism is dead, or it should be! We see a widely popular variant of empiricism in "cognitive science," but then most leaders in this dogma have their origins as behavioral psychologists. One does not easily change one's epistemological stripes. This, I have argued (Novak, 1969; Novak, 1977b, Chapter 2), has been an albatross around the neck of those in educational inquiry and curriculum design. A productive concern with philosophical issues as they relate to educational research has been appearing in recent issues of *Educational Researcher* (see Phillips, 1983).

It is not surprising that many textbooks, lectures, and examinations proceed as if ultimate truths have been found—many teachers and researchers are still at least "closet" empiricists. As some recent research shows (Posner & Strike, 1982; Waterman, 1982), empiricism is still the widely prevailing epistemology. Constructivist views that knowledge is synthesized, modified, and "evolutionary" in character are gaining ground and essentially all contemporary philosophers are some variety of constructivist (Brown, 1979; Popper, 1982). Fortunately, as I see it, constructivist views are also highly compatible with and complementary to an Ausubelian psychology of learning. Since the creation of new knowledge is a *learning* phenomenon on the part of the creator, we should expect congruence between a valid epistemology and a valid psychology of learning.

A decade and a half of attempting to help our colleagues in science departments to improve laboratory instruction gave rise to new educational research studies and methodologies for "unpacking" knowledge from original sources. Out of these efforts, Gowin (1981) came up with what we are finding to be a powerful heuristic for students and researchers, the Epistemological Vee. Figure 12.2 shows the general form and an example of Gowin's Vee applied to physics. Bacon may turn over in his grave to see that the Vee indicates *all* methodological elements on the right side are influenced by our concepts, principles, theories, and philosophies on the left side, but this is obviously the case in examination of any comprehensive report of an inquiry (Watson's, 1968, book is a classic case here). It is also obvious that the left side controls the kinds of questions we choose to ask and the kind of objects

(a)

GOWIN'S EPISTEMOLOGICAL V

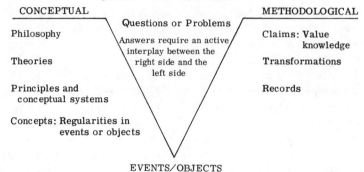

CONCEPTUAL   METHODOLOGICAL

Questions or Problems

Philosophy   Answers require an active   Claims: Value knowledge
                  interplay between the

Theories   right side and the   Transformations
                  left side

Principles and   Records
conceptual systems

Concepts: Regularities in
         events or objects

EVENTS/OBJECTS

(b) Focus Question:
Does each resistor (common resistor, light bulb, NTC, and LDR resistor) follow Ohm's Law?

CONCEPTUAL	METHODOLOGICAL
Philosophy:	**Value claims:**
Scientific knowledge about nature lies in observation and experiment based on theories that organize our facts, reasoning, deepening our understanding.	The whole experiment allows the training of some basic laboratory skills. The knowledge claims show that Ohm's Law is a very particular law.
Theory:	**Knowledge claims:**
Theory of electrodynamics.	The resistor is (is not) linear and the temperature is (is not) constant; therefore, the resistor follows (does not follow) Ohm's Law.
Principles:	
Ohm's Law: $R = \dfrac{V}{I}$ = constant when the temperature T is constant.	**Interpretations:**
	Graphical analysis (R is or is not linear). If R is linear and T is constant, the resistor follows Olm's Law.
Key concepts:	
Electric potential or electric potential difference, electric resistance, electric current, temperature, electric field, light, resistivity, electric charge, energy, conductivity, heat, time, power, conductor, resistor, work, force, electric polarization, current density, photon, temperature coefficient of resistivity, speed, metallic conductor, ion, and linearity.	**Transformations:**
	Tables, V vs. I graphs, calculations of $R = V/I$ and of average R.
	Observations:
	Changes of temperature T and/or luminosity.
	Measurements:
	10 records of electric potential differences and the corresponding electric currents across each resistor. Direct measurements of electric resistances R with a ohmmeter.

INTERPLAY

Event:

An electric potential difference (V) applied on a resistor (common, or light bulb, or NTC, or LDR) establishes a direct electric current I.

or events we choose to observe; empiricism, as noted earlier, has locked psychologists into an elegant but largely fruitless enterprise by constraining the kinds of events observed and questions asked.

One may quarrel with the form of Gowin's Vee heuristic but to argue that it misrepresents contemporary epistemology is more difficult. We have seen no such arguments to date. Most of our students and scholarly colleagues in a wide variety of disciplines have reported profit and usefulness in applying the Vee to inquiries in their disciplines. In any case, the Vee heuristic appears to have sufficient validity for use as a device to help students learn about knowledge and knowledge production, and we have proceeded to use it to this end.

EMPIRICAL STUDIES

My thesis research (Novak, 1958) and our early research studies were based on a "cybernetic" model for human learning derived from the work of Wiener (1948; 1954), Estes (1950) and others.[3] The central focus in these studies was on the nature of problem solving and evaluation of instructional strategies to improve students' problem solving abilities. By 1964, it had become increasingly evident that (1) our data were not consistent with predictions from a cybernetic learning model, and (2) acquisition of specifically relevant knowledge was the key factor in enhanced problem solving performance. These early studies are described elsewhere (Novak, 1977b, Chapters 4 and 8).

Fortunately, Ausubel's *Psychology of Meaningful Learning* became available just about the time our research group was ready to jettison cybernetic theory as a basis for our program. My conviction that any research program must have a theoretical foundation (acquired in part through my work in the sciences and history of science) was enhanced by Ausubel's work, since his theory of meaningful learning not only gave meaning to our earlier research data but also helped to combine my new interests in the central role that concepts play in science learning (Novak, 1964). This change in paradigm guiding our research led to many of the kinds of changes described by Kuhn (1962), including changes in the kinds of research questions we were asking.

---

[3]Wiener's theory of cybernetics describes a model for information storage and processing that had led to modern automation devices and computers. Current information processing and cognitive science are based in part on cybernetic principles.

FIGURE 12.2   Gowin's Vee heuristic showing (a) the 10 basic elements involved in the construction of knowledge, and (b) an example of the use of the Vee as applied in analysis of a physics experiment.

TABLE 12.1   Average Scores and t Values for Experimental and Control Groups on Three Concept Mapping Criteria at Three Intervals in the Semester[a]

Criterion	Group	N	Test 1[b]		N	Test 2[c]		N	Test 3[c]	
Identification of	Experimental	37	1.11	—	35	2.11	—	35	2.46	—
general concepts	Control	35	1.23	−1.23	34	1.53	4.41	33	1.48	7.15
Identification of in-	Experimental	37	1.03	—	35	1.60	—	35	1.89	—
termediate con-	Control	35	1.06	−.63	34	1.18	3.67	33	1.30	4.18
cepts										
Identification of	Experimental	37	1.11	—	35	1.83	—	35	1.97	—
most specific	Control	35	1.20	−.95	34	1.44	2.53	33	1.52	2.81
concepts										

[a]From Moreira, 1977.
[b]$p < .05$.
[c]$p < .01$.

For the next decade, our research groups at Purdue University and, since 1967, at Cornell University sought to apply Ausubelian learning principles in new instructional programs, including the development of audio-tutorial programs at the college, secondary, and elementary school levels (Postlethwait, Novak, & Murray, 1964; 1972). The evaluation studies associated with these instructional programs provided strong support for meaningful learning theory (and considerable negative evidence for the highly popular Piagetian developmental theory). What remained a troublesome problem was the assessment of students' conceptual knowledge and changes in this knowledge as a result of new instruction. We found modified Piagetian clinical interviews (Pines, Novak, Posner, & Van Kirk, 1978) to be useful— but also time consuming and impracticable for classroom teachers. Finally, with the work of Rowell in 1974–1976, we developed an evaluation strategy we call *concept mapping*, and subsequently began to use this strategy not only to analyze interview data but also as a direct instructional technique with students.

Moreira (1977) modified a syllabus for a college physics course on electricity and magnetism to place central emphasis on the key concepts associated with Maxwell's equations. A control group used a more traditional syllabus (based on Halliday & Resnick, 1966) and the experimental group used the modified syllabus. Both groups were taught in a Keller Plan (1968) format and were instructed in making concept maps. Moreira found that the experimental group showed some decline in physics test scores early in the semester, but improved later in the semester (see Table 12.1). Using word association tests and concept maps for evaluation, Moreira found that the experimental group was better able to integrate key physics concepts in

TABLE 12.2   The Number of People Who Used an Entropy
Statement in the Post-Test to Explain the Relationship Between Two
Other Concepts[a]

Relationship	Control	Experimental
Temperature–solubility	2 (2)	8 (18)
Temperature–reaction rate	1 (1)	1 (3)
Temperature–equilibrium	1 (1)	3 (8)
Reaction rate–equilibrium	0	1 (1)
Equilibrium–energy	0	3 (4)

[a]From Cullen, 1983. The numbers in parentheses indicate the total
number of statements that were made.

electricity and magnetism, and to show Maxwell's equations with proper
concept map linkages. The results provided modest evidence for enhanced
meaningful learning of electricity by the experimental students, but not at a
level with substantial practical consequence.

In a related study in college chemistry conducted in 1977, Cullen (1983)
compared performance of students provided with a written study guide
emphasizing the explanatory power of the concept of entropy with students
receiving more traditional study guide materials. Both experimental and
control groups received the same lectures and did the same laboratory experiments. The lecturer discussed the entropy concept as it applied in various topics in chemistry. Using word association evaluation patterned after
Preece (1976), Cullen found that the experimental groups included entropy
more frequently as an associated word on students' word lists for the concepts energy, atom, equilibrium, motion, reaction rate, solubility, and temperature. He also asked students to generate statements using two concept
words provided. Cullen constructed a master list of statements and found
that most student responses could be matched with statements in this list.
Experimental students used the entropy concept more frequently in their
statements than control students (see Table 12.2).

Another finding in Cullen's study was that students who used the entropy
concept to explain novel problems in chemistry were not necessarily the best
students as indicated by course grade (which is based largely on lecture
examination grades). Some 85% of students receive grades of A, B, or C, but
these students were almost randomly distributed when categorized according to the quality of their responses to novel problems. To the extent that
Cullen's evaluation criteria permitted him to identify students with the best
integrated conceptual framework for chemistry, course grades were essentially unrelated to such hierarchically organized knowledge. This finding is
consistent with recent studies that show successful students in secondary or

college science frequently have incomplete or erroneous knowledge of basic phenomena (Gunstone & White, 1981; Champagne, Gunstone, & Klopfer, Chapter 5 & 11, this volume).

Both the Moreira and Cullen studies showed two phenomena that we now find rather consistently in our research. First, typical course evaluation methods do not require use of high order, broad, explanatory concepts and/or integrated frameworks of conceptual knowledge. As a consequence, when course examinations are the criterion of achievement, experimental approaches that emphasize use of broad, explanatory concepts and integration of concept meanings will produce little or no improvement in performance. Second, this kind of experimental approach frequently results in some initial *decline* on conventional course evaluation measures as students struggle to modify their learning patterns from more-rote-mode to more-meaningful-mode approaches. By the end of one semester, experimental subjects may equal or exceed their control counterparts, but achievement based on total test points scored may show no significant differences.

Unfortunately, we have not yet found an instructional setting in which our experimental approaches can be applied over several terms or years in sequentially related courses. Therefore, we can only speculate on the potential contribution of strategies to enhance meaningful learning as measured by either conventional course exams or more cognitively demanding evaluation techniques over the span of secondary and/or college instruction.

In another study relevant to the latter issue, Atkin (1977) found that strategies to enhance meaningful learning in a college organic chemistry course led to no significant differences on knowledge-type test items but a significant gain over control subjects on problem solving items ($\bar{x} = 57.4$ vs. $\bar{x} = 37.7$; $p < .005$). However, on problems of a more sophisticated type, no significant differences were found except that students in both groups who were categorized on a pretest as "knowledge integrators" did much better than their "non-integrator" classmates ($\bar{x} = 19.1$ vs. $\bar{x} = 1.8$; $p < .001$). It is to be expected that several weeks exposure to techniques to enhance meaningful (integrative) learning will not transform a student's total cognitive structure, a large part of which has been acquired over a 20 year span of learning.

In 1974, I began using a draft of *A Theory of Education* as a textbook in my classes. My students, especially my undergraduate students, found that the book was interesting in terms of the issues discussed but what was of most value to them was that it helped them to "learn how to learn." After hearing this from students for several semesters, it gradually sunk into my head that explicit instruction in "learning to learn" would be a sensible thing to do. The studies by Moreira, Cullen, and Atkin were efforts in this direction, but our first explicit efforts to use concept mapping as an instructional tool (rather than an evaluation device) were with college genetics students. We had been using concept mapping with graduate students and faculty in seminars on college teaching, so we knew this strategy was teachable.

Bogden (1977) taught college genetics students the technique of concept mapping and then provided them with concept maps for each lecture given in the course. These maps were subsequently used by Bogden in his discussion sections. His study was not experimental but rather provided an assessment of the feasibility of using concept maps to help students see conceptual relationships and relationships between specific genetic experiments and concepts illustrated or derived from these experiments. Bogden found a mixed reaction from the students, with approximately half of the students reporting some enthusiasm and perceived value in concept maps, and the other half being neutral or negative on their use. Some of the suggestions that grew out of his work were that concept maps should be kept simple (some maps were overwhelmingly comprehensive), lines connecting concepts should be labeled, and students benefit most from concept maps *they* construct. We also found that concept maps helped the professor organize lectures, prepare exams, and construct scoring keys. Some students were highly enthusiastic in regard to the technique.

Partly from the work with Professor Brotman in the genetics course and Professor Holcomb in physics, my colleague Gowin invented a simple heuristic device to illustrate how scientists construct knowledge from the observation of events and objects. He devised his Vee heuristic (Figure 12.2) in November 1977, and we have been using it in essentially its original form since them. The Vee, when used in conjunction with concept mapping, permitted us to offer both metalearning and metaknowledge strategies to our students. Again, piloting the Vee heuristic in our seminars on college teaching, and informally with secondary and college students, we saw that this strategy was teachable. College students frequently reported that they wished they had been taught these ideas earlier, that is, in high school.

Our first systematic effort to introduce metalearning and metaknowledge instruction at the secondary level was begun in 1978. With a grant from the National Science Foundation, we began work with junior high school science classes in the Ithaca area. Initially we planned to use project staff personnel to instruct the students, but it soom became evident that the classroom teacher could introduce and supervise the use of concept mapping and Vee mapping strategies, with some follow-up aid from project staff. However, we did find many teachers were reluctant to engage in the program. The typical junior high school science class is so preoccupied with memorization of technical terms, "cook book" laboratory activities (where they exist), and repeated testing of factual recall that strategies to encourage meaningful learning are generally seen as a diversion from the regular business of the

TABLE 12.3 Acquisition of Concept-Mapping Proficiency
on Five Criteria for Seventh-Grade Students

Criteria	Mean (%)	Standard error of mean
Relationships	80.27	4.89
Hierarchy	105.58	7.21
Branching	71.75	8.87
General to specific	81.11	3.81
Cross links	22.22	5.44
	77.53	4.93

class. What we found in most classrooms was a program designed to achieve near rote-mode learning and little or no desire to deviate from teaching the basic facts.

Gowin (1981) emphasizes that there is a widely held belief among teachers and the general public that teaching *causes* learning. We saw numerous examples of this misconception; and hence our strategies, which serve to help teachers and students *negotiate the meanings* of the material under study, were regarded by some teachers as irrelevant at best and counterproductive at worst. In much of our current research, we are impressed by how widely both teachers and students believe that good teaching *causes* students to learn. We have some distance to go to overcome this misconception and to substitute the recognition that *learning is a responsibility that cannot be shared*—it must be consciously pursued by the student. However, the teacher has a responsibility to help the student grasp the meaning of the material studied so that she or he can choose (or choose not) to learn it.

Over the span of the 1978–1979 school year, we gradually developed strategies for helping teachers to teach concept mapping and Vee mapping. We never did succeed in persuading teachers to accept the primary goal of instruction to be meaningful learning. Virtually all of their class quizzes and semester tests required only verbatim recall of factual information, and none used concept map or Vee map scores as a component of their course grades. In spite of the lack of a grading incentive, however, most students were cooperative and conscientious in preparing and submitting mapping assignments.

We gradually evolved a concept map scoring procedure that was based on five criteria applied to a master map prepared by project staff for the material studied, with student maps scored and recorded as a percentage ratio of student map score divided by master map score. Table 12.3 shows the five criteria used and the mean percentages obtained by a seventh-grade class in the study. These data are typical of scores obtained by both seventh- and eighth-grade classes on a wide variety of science topics. The data show that students do quite well in identifying relationships between concepts (constructing valid propositions) and in constructing an hierarchical organization in their maps (Criteria 2–4). However, they do rather poorly in identifying cross-links, which are propositions linking one section of their concept hierarchy to other sections of the hierarchy. We see cross-links as evidence of integrative reconciliation between related but also distinct conceptual domains. In point of fact, the low performance on this criterion is probably a valid indication that the students' learning was substantially less meaningful than one might hope for.

Additional support for the limited meaningfulness of the students' learning was their performance on a novel problem-solving test that asked the students to write out an explanation of an observed event (such as a cork popping out of a chilled wine bottle after the bottle is left standing in the sun). Students who had used concept mapping and Vee mapping wrote more than twice as many valid propositions to explain this phenomenon when compared with students in regular science classes. Nevertheless, on this winebottle problem, the mappers only wrote an average of 2.61 propositions, whereas some 8 or 10 valid propositions could have been offered on the basis of the materials all classes studied. Thus, we see that either their motivation for writing or the comprehensiveness of their knowledge left more room for improvement. Given the lack of grading incentives for meaningful learning, at least the mappers seemed to be heading in the right direction.

Reviewers of our project proposal submitted to the National Institute of Education–National Science Foundation competition were highly critical in that most believed seventh- and eighth-grade students could not do concept mapping and Vee mapping with any understanding. In fact, seventh graders significantly outperformed eighth graders, though some of this may have resulted from an earlier beginning, hence more weeks of practice in seventh-grade classes. Our data showed continuing gains in mean scores even toward the end of the school year. Clinical interview and anecdotal data clearly show that students understood the use of the strategies, even if they did not fully master them in one school year.

We calculated correlation coefficients between scores on concept maps, Vee maps, class science exams, Scholastic Aptitude Test (SAT) or School and College Ability Test (SCAT) quantitative and verbal scores, and novel problem solving scores. The highest correlations ($r = 0.71–0.78$) were between course exam grades and SAT reading and math, and the lowest correlations were between concept mapping scores and SAT scores ($r = \pm 0.02$). These data indicate that typical course exams (requiring rote recall) measure much of the same aptitudes as SAT tests, but concept maps measure something *substantially different.* It is our judgment that concept map scores are probably more valid indicators of meaningful learning and hence these correlations support the indictment of standardized achievement or ability testing voiced by people such as Gould (1981). Additional data are available in other reports (Novak, Gowin, & Johansen, 1983; Novak & staff, 1981).

In 1980–1981, Gurley, who had been a staff member on our junior high school project, used concept mapping and Vee mapping with half of her high school biology classes in a suburban Chicago public high school. She obtained performance data which were in general highly similar to those obtained wih junior high school students. She also found that student mapping performance continued to improve over the school year. It now seems evident that mapping in only one class for one school year still leaves considerable margin for improvement. Given the rote-learning emphasis in most classes, we should not expect students to master mapping strategies in a limited 10-month experience.

Gurley's department chairman and school counselor interviewed many of her students, and some of the most interesting data in her thesis are these interview transcripts. The following are representative student comments on concept maps and Vee maps.

Responses to questions about the usefulness of concept maping

Concept maps help. I don't like doing them, but they help. I guess I get more out of the reading. When you're doing the concept maps you aren't thinking about what it really means, you're thinking if it makes sense or not.

I can't study from them 'cause I make them too confusing. It's a good idea to concept map 'cause you get more out of what you're reading

Given a choice, well I probably wouldn't do it. I don't like doing 'em, but . . . the map shows out the more important things.

Concept maps used to take all day. Now I also make them from her review tapes.

I study from my maps more than book or tapes or class notes. It brings out *all* the stuff, not just the little words. You see what's in common to the concepts by *how* they connect. It's easier for me now—I get it down in minutes 'cause I've gotten good at looking out for the main things.

I always use my maps. If you just read the book it's different 'cause you might not see the main point of the chapter and how it all fits together. Concept maps are easier to understand. It puts it a different way than the book says it. It gives you the concepts in your own way. They're worth the time—it's easier to learn, for me.

If I don't do well, it's 'cause I don't study, but I love the class—I learn a lot. I may not always get good grades, but I learn a lot.

Concept maps are too confusing. The only good thing is that they bring out the main points and help me remember because I write them down. It's like houses and roads. But I don't think they help the majority of people. The concept maps are harder than questions, but it *does* bring out the points of the chapter in the long run. It's confusing—I still put too many words on the lines. Concept maps take a long time.

Questions would be easier and I'd be happier. Concept maps take time to think out. Biology is hard because of concept maps and essay tests.

Concept maps help me to get the important stuff and just reading alone gets boring. Concept maps put some ideas I wouldn't consider when I'm just reading. I don't mind doing them 'cause it gets the ideas in my head and it makes it not boring. The book is boring. Class is fun.

I can't use concept maps. I'd rather read the chapter over and over. Concept maps are more work. It's different than memorizing—it's all related.

Responses about the usefulness of the Vee

I don't like those—those are dumb. You mark down *why* you do it besides what you do. That is what's different than the lab questions. I guess you really have to understand what you're doing to do the lab.

Occasionally I use it to study for the test. You get more out of lab with the Vee, especially if it's something I don't know well 'cause I'm aware of *why* I'm doing it and the stuff I know already that applies. That's the left side. You get the hang of it after you do a couple. It's better than concept maps.

I don't like Vees. I'd rather have lab questions. They're easy. I guess you understand what you're doing better with a Vee.

You might not ask yourself the main question for the whole lab and you might not know the answer either if you didn't do the Vee. If you don't make yourself remember stuff on the left side, well, it reinforces what you have to know to solve the lab. I always look at all my Vees before a test for that chapter. It usually has to do with a main concept you learn in the chapter. They're useful . . . I like Vees. It helps with the experiment. Puts it in a form that *you* are writing and *you* can understand.

Yeah, Vees help us—an overall summary of everything that's background on it and you need to know. I think Vees help more than concept maps. I'd rather do Vees. They take time, too, but, I mean, they're good.

You do the left side so you know what you're talking about. You have to know your terminology first. I use Vees to go over my prinicples and concepts to make sure I can answer the focus question before a test. Vees are like a study guide for the lab. Vees aren't hard so it doesn't bother me.

If you didn't do the left side of the Vee you wouldn't get *why* you're doing the "doing" side. It explains why and what principles and concepts you are using to understand what happens. I don't mind them, they're a breeze to do.

The Vee lets you figure out what you are doing and you know, you *realize* it, not just like a drone doing it without know what. Vees are easy to do.

The left of the Vee is what you need to get your answer. I like old fashioned lab reports. I don't like Vees—they're too much work.

Taken all together, the students' comments nicely support our contention that learning is the student's responsibility. They also show that students recognize that meaningful learning is hard work and seldom required in most classes. Needless to say, not all students rush to embrace strategies that require meaningful learning.

Since 1980, virtually all our research studies have incorporated concept mapping and/or Vee mapping as strategies. Some of these studies have been directed primarily at questions of instructional design (Buchweitz, 1981; Kinigstein, 1981; Mollura, 1979), whereas others have been more directed at helping students "learn how to learn" (Melby-Robb, 1982; Symington & Novak, 1982). Kahle and her colleagues at Purdue University are completing a related research study with predominantly minority high school students.

The results from these and other studies show promise for the use of the strategies but also serve to illustrate that it will not be easy to accomplish all of the educational changes that are needed. Whenever we use the strategies to assess curriculum, we find serious conceptual gaps or lack of explicit linkages between concepts, poor integration between events or objects presented and concepts, principles and theories needed to interpret observations of the events or objects, and little or no guidance to the student as to significant salient concepts versus peripheral or incidental concepts. Wherever the strategies are applied to student instruction, we find that students have a poor understanding of the nature of knowledge (see, e.g., Waterman, 1982) and often little or no awareness of the difference between essentially rote as contrasted to meaningful learning strategies. It seems obvious that most students are unaware of the nature of their misconceptions or how to correct them.

SUMMARY AND RECOMMENDATIONS

On the basis of our research and other recent relevant studies, we now see significant promise for concept mapping and Vee mapping strategies to help students learn how to learn and to acquire knowledge about knowledge. We see these strategies as holding promise for helping students to understand both the nature and sources of valid as well as invalid conceptions of events or objects. They may in time permit students to gain facility in assessing the power and validity of their idiosyncratic conceptual frameworks. We believe computers can be helpful in this process, and we are currently designing microcomputer programs to help students acquire and use concept mapping and Vee mapping strategies.

At this time, we have only a few strands of evidence, largely anecdotal, that suggest metalearning and metaknowledge will help students to recognize and correct some of their misconceptions. We have some studies underway that should provide some hard data on this question. Our work is largely in the natural sciences, but one of our doctoral students, John Volmink, is getting some promising results with mathematics students. Other students are using mapping strategies in social sciences and humanities.

But there is more that is needed. Schwab (1973) identified four "commonplaces" involved in education: teacher, student, curriculum, and governance. There is still a need to help students learn how to be students and to help teachers learn how to be teachers. Concept mapping and Vee mapping can be helpful, but they are not sufficient for the improvement of teaching and learning. They can also be helpful in curriculum improvement, as some of our research has shown. Gowin (1981) described governance as the control of the *meaning* of experience and while concept mapping and Vee mapping help to shift some of this control to the student, radical changes in school organization will be needed to achieve more liberating meaning in student learning experiences. We believe the theoretical foundations and the methodologies developed so far are on the right track, and we urge and welcome their critical application by other educators. Novak and Gowin (1984) provide guidelines for such application.

What is desperately needed is a study of changes in students' knowledge structures and attitudes toward learning as a result of several years experience in several or all of their classes where *meaningful* leaning is the explicit goal. We are not aware of any such schools. One of our long-term programmatic goals is to help develop such a school and to work with the staff to refine pedagogical and learning strategies as well as evaluation strategies. We believe concept mapping and Vee mapping would play a key role in any school committed to meaningful learning.

REFERENCES

Atkin, J. A. (1977). *An information processing model of learning and problem solving.* Unpublished doctoral dissertation, Cornell University, Ithaca, NY.

Ausubel, D. P. (1963). *The psychology of meaningful verbal learning.* New York: Grune and Stratton.

Ausubel, D. P. (1968). *Educational psychology: A cognitive view.* New York: Holt, Rinehart and Winston.

Ausubel, D. P., Novak, J. D., & Hanesian, H. (1978). *Educational psychology: A cognitive view* (2nd Ed.). New York: Holt, Rinehart and Winston.

Bacon, Sir Francis. *Advancement of learning, novum organum, and new Atlantis.* Chicago: William Benton. Encyclopaedia Britannica, Inc. 1952 (Original work published 1620)

Bogden, C. A. (1977). *The use of concept mapping as a possible strategy for instructional design and evaluation in college genetics.* Unpublished master's thesis, Cornell University, Ithaca, NY.

Brown, H. I. (1979). *Perception, theory and commitment: The new philosophy of science,* Phoenix Edition. Chicago: University of Chicago Press.

Buchweitz, B. (1981). *An epistemological analysis of curriculum and an assessment of concept learning in physics laboratory.* Unpublished doctoral dissertation, Cornell University, Ithaca, NY.

Cordemone, P. F. (1975). *Concept mapping: A technique of analyzing a discipline and its use in the curriculum and instruction in a portion of a college level mathematics skill course.* Unpublished master's thesis, Cornell University, Ithaca, NY.

Cullen, J. (1983). *Concept learning and problem solving: The use of the entropy concept in college teaching.* Unpublished doctoral dissertation, Cornell University, Ithaca, NY.

Estes, W. K. (1950). Toward a statistical theory of learning. *Psychological Review, 57,* 94–107.

Gould, S. J. (1981). *The mismeasure of man.* New York: Norton.

Gowin, D. B. (1981). *Educating.* Ithaca, NY: Cornell University Press.

Gunstone, R. F., & White, R. T. (1981). Understanding of gravity. *Science Education, 65*(3), 291–299.

Gurley, L. I. (1982). *Use of Gowin's vee and concept mapping strategies to teach responsibility for learning in high school biological sciences.* Unpublished doctoral dissertation, Cornell University, Ithaca, NY.

Halliday, D., & Resnick, R. (1966). *Physics.* New York: Wiley.

Helm, H., & Novak, J. D. (1983, June 20–22). *Proceedings of the international seminar on misconceptions in science and mathematics.* Ithaca, NY: Department of Education, Cornell University.

Keller, F. A. (1968). Goodbye teacher. *Journal of Applied Behavioral Analysis* (1), 79–89.

Kiel, F. C. (1979). *Semantic and conceptual development: An ontological perspective.* Cambridge, MA: Harvard University Press.

Kinigstein, J. B. (1981). *A conceptual approach to planning an environmental education curriculum.* Unpublished master's thesis, Cornell University, Ithaca, NY.

Kuhn, T. S. (1962). The structure of scientific revolutions. *International encyclopaedia of unified science, Second Edition. Enlarged Vols. 1 and 2: Foundations of the Unity of Science* (Vol. 2, No. 2.). Chicago: University of Chicago Press.

Levandowski, C. E. (1981). *Epistemology of a physics laboratory on electricity and magnetism.* Unpublished Ph.D. dissertation, Cornell University, Ithaca, NY.

Macnamara, J. (1982). *Names for things: A study of human learning.* Cambridge, MA: MIT Press.

Mayer, R. E. (1981). *The promise of cognitive psychology.* San Francisco: Freeman.

Mayer, R. E. (1983). What have we learned about increasing the meaningfulness of science prose? *Science Education, 67*(2), 223–237.

Melby-Robb, S. J. (1982). *An exploration of the uses of concept mapping with science students labeled low achievers.* Unpublished master's thesis, Cornell University, Ithaca, NY.

Mollura, M. F. (Sister). (1979). *A conceptually-structured curriculum for teaching physiology.* Unpublished doctoral dissertation, Cornell University, Ithaca, NY.

Moreira, M. (1977). *An Ausubelian approach to physics instruction: An experiment in an introductory college course in electromagnetism.* Unpublished doctoral dissertation, Cornell University, Ithaca, NY.

Novak, J. D. (1958). An experimental comparison of a conventional and a project centered method of teaching a college general botany course. *Journal of Experimental Education, 26,* 217–230.

Novak, J. D. (1964, October). Importance of conceptual schemes for science teaching. *The Science Teacher, 31*(6), 10.

Novak, J. D. (1969). A case study of curriculum change—Science since PSSC. *School Science and Mathematics, 69,* 374–384.

Novak, J. D. (1977a). An alternative to Piagetian psychology for science and mathematics education. *Science Education, 61*(4), 453–477.

Novak, J. D. (1977b). *A theory of education.* Ithaca, NY: Cornell University Press.

Novak, J. D. (1980, September 17–21). Metholdological issues in investigating learning. In W. F. Archenhold (Ed.), *Cognitive development research in science and mathematics: Proceedings of an international seminar.* (pp. 129–148) Leeds: Centre for Studies in Science Education, School of Education, University of Leeds.

Novak, J. D., & Gowin, D. B. (1984). *Learning How to Learn.* Cambridge, England: Cambridge University Press.

Novak, J. D. (1982). Psychological and Epistemological Alternatives. In S. Modgil & C. Modgil, (Eds.), *Jean Piaget: Consensus and controversy.* London: Holt, Reinhart and Winston. (pp. 331–349).

Novak, J. D., Gowin, D. R., & Johansen, G. T. The use of concept mapping and knowledge vee mapping with junior high school science students. *Science Education, 67*(5), 625–645.

Novak, J. D., & staff. (1981). *The use of concept mapping and Gowin's "v" mapping instructional strategies in junior high school science.* Ithaca, NY: Cornell University, Department of Education.

Pines, A. L., Novak, J. D., Posner, G. J., & Van Kirk, J. (1978). *The clinical interview: A method for evaluating cognitive structure.* (Research Rep. No. 6). Ithaca, NY: Cornell University, Department of Education, College of Agriculture and Life Sciences.

Phillips, D. C. (1983). After the wake: Postpositivistic educational thought. *Educational Researcher, 12*(5), 4–12.

Popper, K. (1982). *Unended quest: An intellectual autobiography.* London: Open Court.

Posner, G., & Strike, K. (1982, October). *Epistemological assumptions of college students: An initial report.* Paper presented to the 13th Annual Convocation of the Northeastern Educational Research Association, Ellenville, NY.

Postlethwait, S. N., Novak, J. D., & Murray, H. T., Jr. (1972). *The audio-tutorial approach to learning* (3rd Ed.). Minneapolis, MN: Burgess.

Preece, P. F. W. (1976). Mapping cognitive structure: A comparison of methods. *Journal of Educational Psychology, 68*(1), 1–8.

Ridley, D. R., & Novak, J. D. (1983). Sex-related differences in high school science and mathematics enrollments: Do they give males a critical head-start toward science and math related careers? *Alberta Journal of Educational Research, 29*(4), 308–318.

Rowell, R. M. (1978). *Concept mapping: Evaluation of children's science concepts following audio-tutorial instruction.* Unpublished doctoral dissertation, Cornell University, Ithaca, NY.

Schwab, J. (1973). The practical: Translation into curriculum. *School Review,*

Symington, D., & Novak, J. D. (1982). Teaching children how to learn. *The Educational Magazine, 39*(5), 13–16.

Toulmin, S. (1972). *Human understanding. Volume 1: The collective use and evaluation of concepts.* Princeton, NJ: Princeton University Press.

Waterman, M. (1982). *College biology students' belief about scientific knowledge: Foundation for study of epistemological commitments in conceptual change.* Unpublished doctoral dissertation, Cornell University, New York.

Watson, J. D. (1968). *The double helix.* New York: Signet.

Wiener, N. (1948). *Cybernetics.* New York: Wiley.

Wiener, N. (1954). *The human use of human beings* (2nd Ed.). Garden City, NY: Doubleday.

Wittrock, M. C. (1974). Learning as a generative process. *Educational Psychologist, 11*, 87–95.

13

A CONCEPTUAL CHANGE VIEW
OF LEARNING AND
UNDERSTANDING

Kenneth A. Strike and George J. Posner

This chapter has two purposes. The first is to sketch what we refer to as a conceptual change theory of learning. In this section of the chapter, we emphasize the epistemological roots of our view of learning. In the second section, we apply our view of learning to the concept of understanding. Here we develop a rather extensive example of how conceptual change may be applied to the analysis of a historical case. The result gives the reader a good overview of the theory and some sense of how it can be applied.

A CONCEPTUAL CHANGE VIEW OF LEARNING

How might we view learning? What does learning have to do with rationality? We see learning as a rational enterprise, and we understand rationality as having to do with the conditions under which a person is or should be willing to change his or her mind. These two major points lead us to a conceptual change theory of learning which we view as the modern alternative to the empiricism, the main tradition in Western philosophy and philosophy of science for the last several centuries, and of psychology during most of the twentieth century.

LEARNING AS A RATIONAL ENTERPRISE

Much of the way we talk and act about education seems to presuppose an image of the student as a retainer of, rather than a processor of, experience and information. We believe that this is untrue. We suggest instead that learning is best thought of as a process of inquiry. We do not mean by this comment to argue for some variant of the notion of discovery learning as though we believed that it was somehow a crime to tell something to a student or that students never really learn something unless they find it out for themselves. Such views are manifestly untrue. In fact, there are serious epistemological problems with discovery views of teaching and learning (Smith & Anderson, 1983; Strike, 1975). What we do mean, however, is that the task of learning is primarily one of relating what one has encountered (regardless of its source) to one's current ideas. The student who learns something is the one who understands a new idea (which requires it to be located in a semantic syntactical network of concepts), is the one who judges its truth value (which requires relating the idea to appropriate standards of evidence), and is the one who can judge its consistency with other ideas (which may require alterations in the overall conceptual organization). To learn an idea in any other way is to acquire a piece of verbal behavior which one emits to a stimulus, rather than to understand an idea which one can employ in an intellectually productive way.

RATIONALITY AS CONCEPTUAL CHANGE

Our suggestion that rationality has to do with the conditions under which people ought to be willing to change their minds has to do with the particular epistemology that we believe to be true and with why we refer to our view as conceptual change theory. Until recently, many philosophers saw rationality as having to do with the relations between a belief or a set of beliefs (such as a scientific theory) and the experiential or experimental evidence for it. Most debates in the philosophy of science concerned when theories should be held to be verified or falsified by the experimental evidence available. We, however, maintain that the rational acceptance of a theory is not so much a matter of whether it is corroborated by the empirical evidence but whether it solves the problems generated by its conceptual context or by its predecessors. Theories are judged by how successfully they solve their appropriate range of intellectual problems. Those problems tend to be set either by some currently accepted conceptions or by the difficulties left by conceptions that have failed (Brown, 1977; Kuhn, 1970; Lakatos, 1970; Toulmin, 1972). Empirical evidence is not irrelevant here, but being rational has to do primarily with how we solve outstanding problems generated by our current beliefs or by how we move from one view to another.

Given that we believe that learning is a rational activity, it follows that we will view learning in much the same fashion as we view rationality. The important questions are the way learners incorporate new conceptions into current cognitive structures, and the way they replace conceptions which have become disfunctional with new ones. We are not claiming that learning theory is isomorphic with philosophy of science. For example, an understanding of how communities of scientists select or reject new conceptions is an important part of understanding conceptual change in scientific communities, but has little to do with individual learning. We are claiming, however, that questions having to do with individual learning have certain generic structural features, whether they concern a scientist struggling with a new idea on the forefront of knowledge or with a child trying to understand elementary concepts about motion.[1]

It follows from these remarks that it is important to have our philosophy of science straight. We suspect that behind most theories of learning there lies an epistemology. Indeed, we believe that the tradition of academic psychology, so far as learning theory is concerned, has been as sterile as we think it has been because it has essentially been an expression of the epistemology of empiricism. We should then make a few observations about what empiricism is and what the alternatives to it are (Strike, 1982a).

EMPIRICIST EPISTEMOLOGY

For current purposes, we shall represent empiricism as holding the following:

1. All knowledge originates in experience. The traditional empiricist motto is "There is nothing in the mind that was not first in the senses."

For traditional empiricist theories (Hume, 1953), this was both a claim about concept formation and about the evidence for our beliefs. More recent versions of empiricism have wished to distinguish between these aspects. We may then appropriately also express empiricism as a theory of evidence.

2. Experience is the sole evidence for our beliefs. Experience is to be linked to our beliefs by subject matter neutral logical rules (sometimes referred to as scientific method) which allow us to decide when experience indicates that we may accept or should reject our beliefs.

3. Knowledge is additive and bottom up. That is, we add to our store of knowledge either by having some new experience, by confirming some new idea, or by being able to describe our current store of experiences in increasingly general ways.

4. Experience is given to us in atoms sometimes referred to as sensations or sense data. Knowing or learning is essentially a matter of linking these sense data into patterns or regularities.

The essential difficulty with empiricism was pointed out by Plato (1949) in the Meno dialogue. There Plato notes that people who knew nothing would

---

[1] In this chapter, we make no distinction between *conceptions* and *ideas* and use the terms interchangeably. When they refer to students' beliefs constructed in the absence or in spite of instruction, we consider them equivalent to Driver's (1983) alternate frameworks and Viennot's (1979) spontaneous theories.

be incapable of learning anything, for they would be incapable of identifying the truth when they saw it. Let us suggest a simple illustration about a man we shall call Jones. Suppose you did not know Jones or anything about him. Suppose also that Jones sent you a letter offering you a great deal of money if only you would meet him in New York's Grand Central Station on Monday. Jones declined, however, to give you any information as to his appearance or precise location. How would you solve the problem of meeting him? Perhaps your first inclination would be to say that the problem is unsolvable. Since you know nothing about him, you would have no criteria by means of which you could recognize him. This response is, of course, precisely Plato's point. A person who was genuinely a tabula rasa would have no way of ever learning anything for he would have no way of thinking about or drawing conclusions on the basis of experience. Of course, if Jones had offered you a lot of money, you might think about the problem a bit more. Perhaps you would ask whether or not you really knew nothing about him. You might, for example, look at his handwriting and see if you could deduce anything about him from how he signed his name. Or you might look at the postmark on his letter to see if it offered any clues.

The moral of this second attempt is also worth pointing out. It is that your ability to learn from an experience is directly related to the quality of the ideas that you are able to bring to judging it. Had you possessed a good theory of handwriting analysis or known a great deal about postmarks, you might have been able to make some progress in discovering Jones.

CONCEPTUAL CHANGE EPISTEMOLOGY

Contemporary philosophy of science has turned these sorts of observations into a rather complex view of how scientific conceptions are acquired and how they change. Its salient feature is to emphasize the role of current conceptions—of the conceptual context—in generating new knowledge. Knowledge does not simply arise from experience. Rather, it arises from the interaction occurring during problem solving between experience and our current conceptions. Some particulars of this view are:

1. Problems are generated by current conceptions. Intellectual problems do not simply emerge from experience. They are, rather, more likely to be the product of a discrepancy between the intellectual expectations generated by our current conceptions and our actual current capacity to explain experience in terms of these conceptions.

2. Solutions to problems are judged by means of current conceptions. Proposed solutions to problems need to do more than simply explain or predict the phenomena. They need to do so in ways that current conceptions

regard as a successful form of explanation and in ways that are consistent with other knowledge.

3. Conceptions are a precondition of experience. Seeing is something we do with ideas as well as senses. We cannot see what we cannot conceive. Moreover, people who approach the world with different conceptions will see it differently.

4. Current conceptions are a product of a history of conceptual development. This conceptual development has included attempts to understand the world and the modification of conceptions in light of their inadequacies. Current conceptions may not be perfect, but they are rarely arbitrary or altogether unreasonable. There is a presumption that current ideas in the scientific community are accepted because they have had some success in accounting for some range of experience, that they have survived over their competition and that they are products of some degree of testing and refinement. This does not guarantee their truth or their continued adequacy. It does, however, make them objects of respect which should not be lightly dismissed (Strike, 1982b).

EPISTEMOLOGICAL PERSPECTIVES ON LEARNING

Recall that we claimed that epistemological views are the foundation of views of learning. Consider, then, how empiricism and conceptual change theory affect our general view of learning (see also Strike & Posner, 1976).

The crucial difference is that empiricism gives us no reason why we should consider current conceptions as relevant to learning. Empiricism implies that conceptions are taken directly from experience and that current conceptions are unnecessary to learning. Views which emphasize conceptual change, however, assume that students' ability to learn and what students learn depend on the conceptions which they can bring to the experience.

Empiricists are also inclined to see learning as additive. Learning is a matter of accumulating experiences, like a squirrel acquires nuts, and building generalizations on them. Conceptual change views, however, are likely to emphasize the transformation of conceptions in the process of learning. New ideas are not merely added to old ones, they interact with them, sometimes requiring the alteration of both.

Our work on conceptual change to date has focused on large-scale conceptual change akin to Kuhnian paradigm shifts or Lakatosian shifts between research programs. We have used the word *accommodation* to refer to such large-scale conceptual changes and the word *assimilation* to refer to those kinds of learning where a major conceptual revision is not required. The following discussion of the conceptual change and of understanding as a conceptual change are more oriented to accommodation than to assimilation. Here, however, it should be noted that we regard the distinction between accommodation and assimilation as a matter of degree. Moreover, attention needs first and foremost to be focused on the central claim. That claim is that new conceptions are understood, judged, acquired, or rejected in a conceptual context. Explaining learning and understanding is primarily a matter of explaining how this conceptual ecology functions for the student.

CONDITIONS FOR CONCEPTUAL CHANGE

Given our primary interest in accommodation, we can begin to develop a theory of conceptual change which describes how a person's current conceptions function in judging new ones by describing the conditions necessary for an accommodation.

As previously described (Posner et al., 1982), the four conditions for an accommodation are as follows:

1. There must be dissatisfaction with existing conceptions. Scientists and students are unlikely to make major conceptual changes until they believe that less radical changes will not work.
2. A new conception must be minimally understood. The individual must be able to grasp how experience can be structured by a new conception sufficiently to explore the possibilities inherent in it.
3. A new conception must appear initially plausible. Any new conception adopted must at least appear to have the capacity to solve the problems generated by its predecessors, and to fit with other knowledge, experience, and help. Otherwise it will not appear a plausible choice.
4. A new conception should suggest the possibility of a fruitful research program. It should have the potential to be extended, to open up new areas of inquiry and to have technological and/or explanatory power.

FEATURES OF A CONCEPTUAL ECOLOGY

As we have discussed previously (Posner et al., 1982), an individual's current cognitive resources, his or her conceptual ecology, will influence the selection of a new conception. The literature in philosophy of science and our own work have suggested that the following kinds of resources are particularly important determinants of the direction of an accommodation.

1. Anomalies. The character of the specific failures of a given idea are an important part of the ecology which selects its successor.
2. Analogies and metaphors. These can serve to suggest new ideas and to make them understandable.

3.   Exemplars and images. Prototypical examples, thought experiments, imagined or artificially simulated objects, and processes all influence a person's intuitive sense of what is reasonable.

4.   Past experience. Conceptions which appear to contradict one's past experience are unlikely to be accepted.

5.   Epistemological commitments.

a.   *Explanatory ideals.* Most fields have some subject matter specific views concerning what counts as a successful explanation in the field.

b.   *General views about the character of knowledge.* Some standards for successful knowledge such as elegance, economy, parsimony, and not being excessively ad hoc seem subject matter neutral.

6.   Metaphysical beliefs and concepts.

a.   *Metaphysical beliefs about science.* Beliefs concerning the extent of orderliness, symmetry, or nonrandomness of the universe are often important in scientific work and can result in epistemological views which, in turn, can select or reject particular kinds of explanations. Beliefs about the relations between science and commonplace experience are also important here.

b.   *Metaphysical concepts of science.* Particular scientific conceptions often have a metaphysical quality in that they are beliefs about the ultimate nature of the universe and are immune from direct empirical refutation. A belief in absolute space or time is an example.

7.   Other knowledge.

a.   *Knowledge in other fields:* New ideas must be compatible with other things people believe to be true.

b.   *Competing conceptions:* One condition for the selection of a new conception is that it should appear to have more promise than its competitors.

TABLE 13.1   The Relationship of Features of a Conceptual Ecology to the Conditions of an Accommodation

Features of a conceptual ecology	Conditions for an accommodation			
	Dissatisfaction	Minimal understanding	Plausibility	Fruitfulness
Anomalies	X		X	X
Analogies and metaphors		X	X	
Exemplars and images		X	X	
Epistemological commitments	X		X	
Metaphysical beliefs and concepts	X		X	
Past experience	X		X	X
Other knowledge	X		X	X

These two dimensions of conceptual change relate directly to each other as summarized in Table 13.1. Let us examine these relationships more closely and thereby explicate a model of conceptual change.

DISSATISFACTION WITH EXISTING CONCEPTIONS

Generally, a new conception is unlikely to displace an old one, unless the old one encounters serious difficulties and a new understandable and initially plausible conception is available that resolves these difficulties. That is, the individual must first view an existing conception with some dissatisfaction before he or she will seriously consider a new one. Dissatisfaction results from the individual experiencing one or more of the following conditions:

1. A conception is incapable of interpreting experiences presumed to be interpretable (resulting in an anomaly).
2. A conception is seen to be no longer necessary in the interpretation of experiences previously considered significant. This may be a consequence of another conception's greater success in interpreting the experiences or conception reducing the significance of the experiences.
3. A conception is incapable of solving some problems that it presumably should be able to solve.
4. A conception violates an epistemological or metaphysical standard.
5. The implications of a conception are unacceptable.
6. A conception becomes inconsistent with knowledge in other areas.

One major source of dissatisfaction is the anomaly. Each time a person unsuccessfully attempts to assimilate an experience or a new conception into his or her existing network of conceptions, that person experiences an anomaly. An anomaly exists when one is unable to assimilate something that is presumed assimilable—one simply cannot make sense of something.

When faced with an anomaly, the individual (scientist or student) has several alternatives. One may come to the conclusion that one's existing conceptions require some fundamental revisions (i.e., an accommodation) in order to eliminate the conflict. But this is the most difficult and, therefore, the most unlikely approach, especially when there are other possibilities:

1. Rejection of the observational theory;
2. A lack of concern with experimental findings on the grounds that they are irrelevant to one's current conception;
3. A compartmentalization of knowledge to prevent the new information from conflicting with existing belief ("Science doesn't have anything to do with the real world"); and

4. An attempt to assimilate the new information into existing conceptions (e.g., "Newtonizing" relativistic phenomena).

This analysis suggests that the presentation of anomalies will produce dissatisfaction with an existing conception only if:

1. Students understand why the experimental finding represents an anomaly;
2. Students believe that it is necessary to reconcile the findings with their existing conceptions;
3. Students are committed to the reduction of inconsistencies among the beliefs they hold (Strike & Posner, 1983); and
4. Attempts to assimilate the findings into the students' existing conceptions are seen not to work.

Given the improbability that all these conditions will be met, it is no wonder that few students find their current conceptions weakened by anomalies. Why consider alternatives to a view they hold when they are unconvinced of the inadequacy of their conceptions?

Minimal Understanding of a New Conception

In order for students to consider an alternative conception, they must understand it at least at a minimal level.

We argue in the second part of this chapter that understanding an idea requires that it be viewed within a context of other ideas. That is, understanding entails finding a niche within a conceptual ecology. In the second part, we distinguish minimal understanding from a fuller understanding of the idea, and understanding from accommodation. For present purposes, it suffices to suggest two requirements for minimal understanding of an idea.

1. It is necessary to construct or identify a framework in which to locate the new idea. Metaphors and analogies enable the student to borrow frameworks from other contexts. Forming images enables students to construct visual frameworks for this purpose.
2. It is necessary to attach the framework to the world in at least prototypical ways. Exemplars are standard cases to which a framework has been applied.

The importance of frameworks in understanding ideas is analogous to the cognitive scientists' concern for representation of knowledge. As cognitive scientists point out, ideas cannot function psychologically unless the student can internally represent them. Representations in the form of images and/or networks of propositions function both passively and actively. They function passively as a format into which information must be fit. In paragraph comprehension tasks, for example, anomalous sentences are confusing (i.e., not understandable) because they cannot be fit into the representations being built and, thus, are not easily entered into the reader's memory (Bransford & Johnson, 1973). Representations also function actively as a plan for directing one's attention and conducting purposeful searches (Neisser, 1976). The inability of readers to remember an anomalous sentence in an otherwise coherent paragraph may be attributed to the readers' inattention to it. Once students can represent a new conception, they can consider its plausibility.

INITIAL PLAUSIBILITY OF A NEW CONCEPTION

Regardless of how understandable one finds a conception, it may still appear counterintuitive. What makes a conception counterintuitive?

Initial plausibility can be thought of as the anticipated degree of fit of a new conception into an existing conceptual ecology. There appear to be at least six ways by which a conception can become initially plausible.

1.   One finds it consistent with one's current metaphysical beliefs and epistemological commitments, that is, one's fundamental assumptions.

2.   One finds the conception to be consistent with other theories or knowledge about which one is aware.

3.   One finds the conception to be consistent with past experience.

4.   One finds or can create images for the conception, which match one's sense of what the world is or could be like.

5.   One finds the new conception capable of solving problems of which one is aware (i.e., resolving anomalies).

6.   One finds the conception to be analogous to some other conception with which he or she is already familiar.

FRUITFULNESS OF A NEW CONCEPTION

A person becomes committed to a conception because it helps interpret experiences, solve problems, and, in certain cases, meet spiritual or emotional needs. A new conception should do more than the prior conception for the person, if it is to be considered fruitful, but it must do so without sacrificing any of the prior conception's benefits, or must provide sufficient incentives for any required sacrifice. It is no wonder, then, that a few anomalous pieces of data cannot shake a person's commitment to a prior conception (e.g., scientific creationism), even though a new conception (e.g., evolution) resolves the anomalies. Assuming a person can believe in a new conception that resolves apparent anomalies while doing as much for the person as the prior conception, then the person may actively attempt to map the new conception onto the world and to extend it. If doing this leads to new insights and discoveries, then the new conception will appear fruitful and the accommodation of it will seem persuasive. A new conception will appear fruitful to the extent that students are aware of, can generate, or can understand novel practical applications or experiments which the new conception suggests.

THE CHARACTER OF ACCOMMODATION

Our description of the four conditions of a successful accommodation may have suggested a fairly straightforward linear process: students' dissatisfaction with an existing conception, followed by finding a new conception understandable, leading to an initial belief in its plausibility, and concluding with the belief that the new conception is ultimately fruitful.

However, it should be clear that this account is oversimplified, since many basic conceptions are so complex that at a particular time one is likely to accommodate certain aspects but not others. We have, of course, described accommodation as a radical change in a person's conceptual system. That an accommodation is a radical change does not, however, entail that it is abrupt. Indeed, there are good reasons to suppose that, for students, accommodation will be a gradual and piecemeal affair. Students are unlikely to have at the outset a clear or well-developed grasp of any given conception and what it entails about the world. For them, accommodation may be a process of taking an initial step toward a new conception by accepting some of its claims and then gradually modifying other ideas, as they more fully realize the meaning and implication of these new commitments. Accommodation, particularly for the novice, may often occur as a gradual adjustment, each new adjustment laying the groundwork for further adjustments but where the end result is a substantial conceptual reorganization or change.

Accommodation can be viewed as a competition between conceptions. Once students are aware of an understandable and initially plausible alternative to an existing conception, the relative status of these conceptions is the issue. Dissatisfaction with the existing conception decreases its status, while exploring the fruitfulness of an alternative conception increases the alternative's status (Hewson, 1983). As long as the existing conception's status is greater than the alternative's (for whatever reason), accommodation will not proceed. Whenever the alternative's status exceeds the existing conception's status, accommodation, for the time being, will move forward. But, as we have discussed, many factors affect status. Therefore, competition between conceptions results in a process of accommodation characterized by temporary advances, frequent retreats, and periods of indecision.

Our research also indicates that what may initially appear as an accommodation may turn out to be something less than that (Posner et al., 1982). Typically, students will attempt various strategies to escape the full implication of a new conception or to reconcile it with existing beliefs. Accommodation may, thus, have to wait until some unfruitful attempts at assimilation are worked through. It rarely seems characterized by either a flash of insight, in which old ideas fall away to be replaced by new visions, or as a steady logical progression from one commitment to another. Rather, it involves much fumbling about and many false starts and mistakes.

CONCEPTUAL CHANGE AND UNDERSTANDING

In this section, we apply our theory of conceptual change by attempting to characterize what it means to understand something and make a few observations about how understanding might be promoted.

MEANING

The key idea to understanding understanding is that of meaning. To understand an idea is to know what it means. Here we use meaning in its linguistic sense. We are not talking about the importance of an idea or the consequences of an idea. The question, rather, is how sounds or squiggles on a page get to be about something and once they have gotten to be about something, how people come to discover what it is that they are about.

Meaning has the following relationship to truth. A proposition need not be true to be meaningful. Clearly we know what false propositions mean. Indeed, it seems as though meaningfulness is presupposed by truth in the sense that only meaningful propositions can be true or false. For the issue of truth to arise, something meaningful must have been said.

In fact, the relationship between meaning and truth may be more intimate than this suggests. Some philosophers have argued that to know what a proposition means is precisely, and nothing more than, to know the conditions under which the proposition would be true (Ayer, 1976; Wittgenstein, 1921/1961). Intuitively there is much to be said for this view. It is prima facie evidence that a person does not know what a proposition means if he or she encounters a situation which ought to confirm or falsify the proposition and fails to recognize that. If you see someone looking at the telephone on your desk and asking, "Is there a telephone on your desk?", you have reasons to wonder if the individual knows what it means for there to be a telephone on your desk.

With a few qualifications (most of which are not noted here), we agree with this view of the relationship between meaning and truth. We do not, however, formulate the idea as it has traditionally been formulated. Our views on meaning can best be expressed by seeing how and why they depart from what we term *the traditional view*.

THE TRADITIONAL VIEW

The traditional view starts with a sharp distinction between semantics and syntax. Syntax has to do with the arrangement of words. Syntactical rules specify such things as what counts as a well-formed sentence, how sentences can be transformed, and whether one sentence can be inferred from others. In the traditional view, questions of syntax can be decided independently of questions of semantics. We can, for example, know if an argument exemplifies a rule of inference or if a sentence is well-formed, merely by inspecting its syntactical structure. We need not know what any of the nonlogical vocabulary mean.

Semantic rules are conceived as connecting words to the world. To know what a particular word means is to know the range of appearances to which the word can correctly be ascribed. To know what the word dog means is to know the range of appearances that are correctly described as dogs.

Semantic rules are likewise seen as independent of syntax. Words mean what they mean independent of the linguistic context in which they occur. Meaningful propositions, then, are seen as constructed by embedding meaningful terms in a syntactically well-formed expression. This view embodies several noteworthy components. The first might be described as linguistic atomism. Meaningful propositions are constructed from meaningful parts. The meaning of the whole is a product of the meaning of its parts.

The second idea, which has been called the verification principle (Ayer, 1976), is that the meaning of a (nonanalytic) proposition is exhaustively specified by an enumeration of the empirical states of affairs which would confirm it. To know what a proposition means is to know how the world would appear if it were true. It is, thus, to know what would count as verifying the proposition.

The third idea is that there is a sharp distinction between analytic and synthetic propositions. Analytic propositions are true in virtue of their syntactical form. We know that $A = A$ by virtue of its syntactical structure. Synthetic propositions are those which are not true in this way. According to this view, a proposition is either clearly analytic or synthetic.

The reader should note that these features of what we have thus far referred to as the traditional view suggest that it has much in common with traditional empiricism. In fact, it is derived from a more recent view of empiricism—one which has augmented the empiricist tradition with the tools of modern symbolic logic. Central commitments of empiricism endure, however. The verification principle and linguistic atomism, for example, are both a reflection of the view that meaning begins in experience and must ultimately be reducible to experience. We no longer have the traditional motto that there is nothing in the mind that was not first in the senses, but the assumption of the verification principle, that all meaningful terms can be shown to be logically equivalent to statements containing only terms which refer to direct experiences, is its modern equivalent. Linguistic atomism and the sharp distinction between analytic and synthetic propositions are the heirs of the atomism of traditional empiricism and reflect the empiricist conviction that experience apart from context is sufficient to grasp the meaning of an idea.

This viewpoint has the considerable merit of being internally consistent and quite elegant. It has the liability of not being true. Among its difficulties are the following. First, what words mean is not independent of their linguistic context. For example, if one says, "I spilled my _____ on the table," we know more about what will fit into the blank than that it will be a noun. We expect it to be something fluid and edible. If the word turns out to be "wisdom," we must do a bit of work to make sense of the sentence. We will have to interpret wisdom so that it can be the sort of thing that can be spilled, or (more plausibly) we will have to reinterpret "spill" so that it is the kind of thing that can be done to wisdom. Meaning is context dependent. We cannot describe the syntax and semantics of a natural language independently of one another. Atomism is false.

Second, there is not a sharp distinction between analytic and synthetic propositions (see Quine, 1961). A proposition may be analytic in one context and synthetic in another. A chemist is likely to treat "water is H_2O" as a definition. It will be true in virtue of his syntactical rules of equivalence. For most of us, "water is H_2O" is an empirical claim requiring evidence. Whether or not all swans are white is neither a conceptual question to be decided by analyzing our language or an empirical question to be decided by the discovery of some new facts. It is a question of how it makes sense to classify black swan-like birds.

Finally, there are numerous propositions which lack any empirical meaning independent of the theoretical context in which they are embedded. The verification principle is not true as a thesis about individual propositions. Propositions imply things about the world in bunches, not one at a time. Thus, it is not reasonable to expect propositions to have an empirical meaning independently of the theoretical context in which they occur. To grasp the meaning of an idea, it is necessary to see how it fits into a conceptual niche.

Moreover, the verification principle is probably not true even of groups of propositions. No one has succeeded in showing that the meaning of any scientific claim or theory can be reduced to its empirical implications. It is generally conceded that the task is impossible.

A CONCEPTUAL CHANGE VIEW

These points suggest the following conclusions.

First, linguistic atomism is false. The meaning of any part is dependent on how it fits into the whole. Grasping the meaning of any particular conception is a matter of seeing how it fits into the whole. To use Wittgenstein's phrase, it is a matter of seeing how it is used, of determining the role it plays in the language game.

Second, propositions have meaning in sets, and have empirical implications in sets. What is meant by a given proposition depends on the theory or the set of assumptions in which it is embedded, that is, on the proposition's conceptual context. Predictions are not derived from single assertions, but from sets of assertions.

Third, if the claim that to grasp the meaning of an idea is to know the conditions under which it will be true is to be defended, we will have to have a different idea about truth conditions than that expressed in the traditional view. The truth or (as we prefer) reasonableness of a proposition will have to depend not only on being able to generate confirmed predictions, but on relationships to other propositions.

Grasping the meaning of an idea is normally a matter of seeing how that idea is interpreted or applied within a certain conceptual context.[2] To try to explain equality, for example, one might proceed by trying to show how equality would be formulated within a liberal democratic point of view. It is a matter of understanding the problems which doctrines about equality are intended to solve, of noting what might count as a solution to such problems, and of coming to appreciate the constraints placed on the idea. Why is it, for example, that liberals tend to be more interested in equal opportunity than in equal results? Understanding the idea is a matter of being able to formulate it within the requirements and constraints of a more general political theory. These requirements and constraints are both the truth conditions of the idea imposed by the theory and the conditions of an understandable articulation of what the idea means. The meaning of the idea cannot be understood apart from its conceptual home in the broader theory. Located in a different theoretical home, equality means something different.

---

[2]Our previous argument against the distinction between analytic and synthetic propositions based on Quine (1961) implies that there is not a sharp distinction between propositions, on the one hand, and conceptions and ideas, on the other hand. Thus, we apply our analysis of the meaning of propositions to a discussion of the meaning of ideas (or conceptions).

AN EXAMPLE

Let us examine a case of understanding from a conceptual change perspective.[3] We will suggest that when Freud introduced his theory of psychoanalysis, he had to produce an accommodation in most of his audience. This was necessary because the conception of mind which was widely held at the time he began his writing was such that most of the things which Freud had to say would have appeared highly counterintuitive and virtually unintelligible to anyone who held it. Freud's theories could not be assimilated without a change of mind.

The essence of the problem was that the prevailing conception of mind (inherited from René Descartes) was that mind is consciousness. Mind is by its very nature the sort of thing of which one is directly aware. Immediate awareness is the essential and defining characteristic of the mental.

To anyone with such a view, Freud's conception of the unconscious is not just a new idea requiring suitable empirical evidence. It is a suggestion that must appear compellingly counterintuitive. It is not just that the person will not see how to fit the idea into some pre-existent framework. Rather, the individual possesses a framework which will reject the logical coherence of Freud's proposal. Freud thus not only faced the task of persuading people of the truth of his views, but of rendering them understandable. How did he proceed?

We analyze his approach according to the general theory of accommodation sketched earlier. In order to succeed, Freud had to persuade people that their current conceptions were inadequate, he had to render the new ideas intelligible, he had to make them appear initially plausible, and he had to show them to be fruitful.

Freud's strategy for creating dissatisfaction with current conceptions was to generate a set of anomalies for the view that mind is consciousness which require his view to deal with them coherently. His basic example was hypnosis, followed closely by his analysis of mistakes. Hypnosis seems to reveal mental content of which people are not aware. Freud also argued persuasively that many mistakes are purposeful and, thus, reveal purposes hidden from those who act on them.

Freud used several devices to make his view of mental activity at least minimally understandable. Perhaps most important was the use of metaphors. Ideas such as repression and censorship may be so familiar to us in their psychoanalytic context that they will hardly seem to be metaphors at

---

[3]Note that while we consider the following reconstruction of Freudian psycholoanalysis as historically plausible, its value as an illustration does not depend on its historicity. Not being Freud scholars, we do not wish to vouch for the historical accuracy of the details of the reconstruction.

all. Their role in Freud's exposition, however, seems to be to provide illustrations of how ideas can be known, yet hidden. They are political metaphors which were perhaps themselves more intelligible to a nineteenth century German audience which was familiar with organic views of the state. Censorship and repression are illustrations of how things which are known can nevertheless be rejected by or hidden from the knower.

Another strategy of Freud's was his emphasis on the symbolic character of mental content. Symbols and codes are another case of things which can be known, yet hidden. An encoded message is a case where we are aware, yet not aware.

It is also important to note that these models and metaphors of something that is known yet hidden are to be understood against the background of a view of mind which maintains a degree of continuity with the Cartesian tradition. Freud did not entirely reject the idea that mind equals consciousness. Mental content is still in principle conscious and, thus, introspectible in principle. The models and metaphors, represented in such ideas as repression, censorship, and symbolism, seem designed to help people understand how mental content could be the kind of thing people are aware of in principle when they seem not to be aware of it in fact.

Freud seems to have used these metaphors to create a new network of ideas in which to locate his idea of the mental. He also had a set of strategies to help people see how to attach the theory to the world, which is often accomplished via what we call exemplars. These are standard cases which exhibit the theory in its application to paradigmatic phenomena. Freud's analysis of dreams and mistakes seem to play this role in his exposition of his theory.

The initial plausibility of Freud's analysis was demonstrated by showing that the anomalies, the phenomena which cannot be dealt with on the assumption that mind is consciousness, can be dealt with if one is willing to assume that there is such a thing as unconscious mental content. The phenomena of hypnosis and mistakes make sense with that assumption and not otherwise.

Fruitfulness was shown first, by showing that the assumption of the unconscious provides a plausible way to deal with the otherwise mysterious phenomena of mental illness. Freud also made elaborate attempts to demonstrate the fruitfulness of the idea of the unconscious by applying psychoanalytic concepts to as diverse a range of human phenomena as religion and art.

The point is that context is necessary to meaning. Ideas become meaningful when they fit into a conceptual niche in a determinate way. When an accommodation is required, the difficulty is that the context in which a new idea is to be understood is uncertain. It may be unclear how a given idea fits into whatever else people believe. Often current ideas contain elements which are incompatible with the new idea. In such a case, the task is to take current ideas and weave them into a new pattern into which the new idea fits. This is what Freud has done. He had constructed a conceptual context in which the idea of the unconscious could make sense. Having inherited a conceptual ecology hostile to the idea of the unconscious, he wove ideas already familiar into a different kind of ecological niche into which the unconscious fitted comfortably. He then was able to show people how this idea could be attached to the world and how it could be used to solve problems.

Understanding and Accommodation

To recapitulate, we began with an argument to the effect that a particular view concerning learning was false. The way in which the view was false suggested that understanding how ideas become meaningful was in large part a matter of seeing how they fit into a conceptual context.

How does this view of understanding fit into the overall theory of conceptual change? In the first section, we claim that one of the conditions for an accommodation was that the new idea be minimally understood. That is, it must make sense to the learner. The present section is, in a sense, an explication of what is involved in making an idea understandable.

What may seem anomalous about this discussion is that it may appear that the idea of understanding has expanded so as to incorporate virtually all of the conditions of conceptual change. However, one should not conclude from this expansion that understanding a new idea is the same as accommodating it. To do so would miss the distinction between understanding and commitment. To claim that one has a thorough understanding of an idea requires that the person see how to employ it to solve anomalies generated by some prior view and how to apply it to other views. It is to know how the conception would relate to other ideas if it were true. The claim does not, however, require agreement with the idea, because it is possible to understand an idea with which one disagrees.

For example, one of the authors has done some work examining Marxist analyses of education. He understands how Marxists apply their analytic framework to educational phenomena. He can construct well-formed Marxist analyses of new events. But he is not a Marxist, because he does not accept the overall framework of Marxist analysis.

Our remarks thus suggest that it is useful to distinguish three different ideas:

1. <u>Minimal understanding</u>. The minimal level required for a person to begin to entertain the possibility of the truth or reasonableness of a new conception. It is having enough sense of the idea to begin to explore its

possibilities and to see some of its implications. An appreciation of the full range of implications of the conception will be lacking at this level. The person who understands at this level will generally be able to apply the conception to simple or stereotyped problems, but will not see how the idea can be used in more complex situations.

2.   Fuller understanding. The individual has a level of understanding of the conception that resembles that of an expert. The person sees a wide range of the implications of the conception and is able to apply it to complex and novel situations.

The point of wishing to distinguish between these levels of understanding can be seen in some of the expert–novice literature (e.g., Larkin, McDermott, Simon, & Simon, 1980) where people show notable differences in their ability to apply a given conception to different sorts of problems. In our own work (Posner et al., 1982) some interviews of college physics students concerning the special theory of relativity suggested the distinction. Students who were able to formulate the basic postulates of special relativity and use them to solve simple problems often lost sight of their implications when given more complex problems. Indeed, they would frequently fall back on Newtonian ideas. Such phenomena suggest that it is worth distinguishing between minimal and fuller understanding of a conception.

3.   Accommodation   Accommodation involves not only understanding, but a degree of acceptance. In order to claim that a person has accommodated a new conception, a person must have at least a minimal grasp of the meaning of it. One cannot believe something one does not understand. However, the claim that an accommodation has occurred adds to the idea of understanding the notion of commitment or belief. The conception is accepted. Note that while generally full understanding of a conception is not required in order to accept it, it is also the case that an accommodation which involves only a minimal level of understanding is insecure. Students may abandon such a commitment once they see the full range of its implications, or they may unwittingly use inappropriate ideas to analyze complex situations to which the new conception should apply. In some cases, minimal understanding can make it difficult to decide whether an accommodation has really occurred.

CONCLUSION

Many pedagogical implications can be drawn from these ideas. However, it is not our purpose to design a pedagogical theory in these pages. Rather we have sought to sketch the general features of a view of how to think about learning—the fundamental assumptions of a research program, if you will. By extending the account to the concept of understanding, we have tried to show that the ideas have the flexibility to illuminate a wide range of phenomena of interest to educators. The theory has not touched all bases. It has not been integrated with the affective aspects of learning (see Strike & Posner, 1983), nor with habit formation or motor learning. For those of you among the readers whose primary interests are in the learning of subject matter, we would commend the idea that learning is conceptual change as a fruitful approach.

REFERENCES

Ayer, A. J. (1976). *Language, truth and logic.* New York: Dover.
Bransford, J. D., & Johnson, M. K. (1973). Considerations of some problems of comprehension. In W. G. Chase (Ed.), *Visual information processing* (pp. 383–438) New York: Academic Press.
Brown, H. I. (1977). *Perception, theory and commitment: The new philosophy of science.* Chicago: University of Chicago Press.
Hewson, P. W. (1983, April). *Microcomputers and conceptual change: The use of a microcomputer program to diagnose and remediate an alternative conception of speed.* Paper presented at the annual meeting of the American Educational Research Association, Montreal.
Hume, D. (1955). *An inquiry concerning human understanding* New York: Liberal Arts Press.
Kuhn, T. S. (1970). *The structure of scientific revolutions* (2nd ed.). Chicago: University of Chicago Press.
Lakatos, I. (1970). Falsification and the methodology of scientific research programs. In I. Lakatos and A. Musgrave (Eds.), *Criticism and the growth of knowledge* (pp. 91–196). Cambridge: Cambridge University Press.
Larkin, J., McDermott, L., Simon, D. P., & Simon, H. A. (1980). Expert and novice performance in solving physics problems. *Science, 208,* 1335.
Neisser, U. (1976). *Cognition and reality.* San Francisco: W. H. Freeman.
Plato, V. (1949). *Meno* [Library of Liberal Arts]. (B. Jewett, Trans.). Indianapolis: Bobbs-Merrill.
Posner, G. J., Strike, K. A., Hewson, P. W., & Gertzog, W. A. (1982). Accommodation of a scientific conception: Toward a theory of conceptual change. *Science Education, 66*(2), 211–227.
Quine, W. V. O. (1961). Two dogmas of empiricism. In W. V. O. Quine (Ed.), *From a logical point of view* (pp. 20–46). Cambridge: Harvard University Press.
Smith, E. L., & Anderson, C. W. (1983, April). *The effects of teacher's guides on teacher planning and classroom instruction in activity-based science.* Paper presented at the annual meeting of the American Educational Research Association, Montreal.
Strike, K. A. (1975). The logic of learning by discovery. *Review of Educational Research, 45*(3), 461–483.
Strike, K. A. (1982a). *Education policy and the just society.* Urbana, IL: University of Illinois Press.
Strike, K. A. (1982b). *Liberty and learning.* Oxford: Martin Robertson.
Strike, K. A., & Posner, G. J. (1976). Epistemological perspectives on conceptions of curriculum organization and learning. In L. S. Shulman (Ed.), *Review of research in education* (Vol. 4, pp. 106–141). Itasca, IL: F. E. Peacock.
Strike, K. A., & Posner, G. J. (1983, April). *Understanding from a conceptual change point of*

view. Paper presented at the annual meeting of the American Educational Research Association, Montreal.

Toulmin, S. (1972). *Human understanding.* Princeton, NJ: Princeton University Press.

Viennot, L. (1979). Spontaneous reasoning in elementary dynamics. *European Journal of Science Education, 1*(2), 205–222.

Wittgenstein, L. (1961). *Tractatus logico-philosophicus* (D. F. Peers and B. F. McGuiness, Trans.) London: Routledge and Kegan Paul. (Original work published 1921)

AN APPROACH TO DESCRIBING LEARNING AS CHANGE BETWEEN QUALITATIVELY DIFFERENT CONCEPTIONS

Bengt Johansson, Ference Marton, and Lennart Svensson

Our intention here is simply to describe an approach to the study of learning, developed in more than a decade by our research group. We describe our methodological stance without making explicit comparisons with other approaches. We illustrate our way of reasoning by means of a concrete example and general principles. Because we believe that the research methodology always has to be a function of our conceptualization of the phenomenon studied, in the first part of this chapter, we outline our way of seeing learning, and in the second part, we elaborate its methodological implications.

A VIEW OF LEARNING

Assumptions about School Learning

One of the possible answers to the question, "What do students learn in school?" is that some of them learn to solve certain types of problems in certain subject areas. What types of problems they learn to solve and to what extent can be clarified by means of systematic investigation.

Consider the following problem:

> In an experiment with a ball, it is found that when the ball falls, it is affected by the air with a braking force F, which is proportional to the velocity of the ball, v, i.e., $F = k \cdot v$, where k in this case is 0.32 Ns/m. What would the final velocity of the ball be if it were dropped from a high altitude? The ball's mass is 0.20 kg.

This task appeared in the annual nationwide achievement test in physics for the second year of the science and technology line of the Swedish upper-secondary school (equivalent to senior high school). The normal age of the students at this level is 18 years, and it is their eleventh school year.

Realizing that after a phase of accelerated motion, the magnitude of gravitational force affecting the ball ($F_g = mg$) will equal the magnitude of the force of air resistance ($F = kv$), the following equation can be set up and solved for v:

$$kv = mg, \text{ where } k = 0.32 \text{ Ns/m}$$
$$m = 0.20 \text{ kg}$$
$$g = 9.81 \text{ m/sec}^2.$$

In order to gain a point for this problem, the students needed only to write the correct value (6.1 m/s) on the answer sheet. About 70% of them did; it is fairly easy problem. In order to solve this problem, one has to find the right formula, set up the equation, replace the symbols with the numerical values, and carry out the calculation. Solving this problem should not cause any great trouble, especially if one had faced a similar problem previously. Thus, this is one way of answering the question, "What is learned?" and it is correct in one sense. However, this way of answering the question is based implicitly on a certain view of learning, and we have found both the accepted answer and the view of learning too restricted. Our approach to answering this question is different, and it is different because our view of learning is different.

If we subject the preceding problem to a somewhat closer examination, we may discover that its solution is based on a certain way of viewing the phenomenon in question: a certain conception of bodies moving at a constant velocity is presupposed. Underlying the relation between the gravitational force and the air resistance is the notion that a body continues to move at a constant velocity when the resultant of the forces acting on it is zero.

This notion is somewhat counterintuitive and it is surely not a part of most of the students' thinking when they begin their studies in secondary school. For many, it is not a part of their thinking after they have completed their studies, even if they have learned to solve problems by using that very notion.

Meaningfully incorporating a concept into one's thinking process means a change in the way one apprehends certain phenomena in the surrounding world. It is this kind of change that corresponds to our view of the essence of learning. This kind of change represents the type of learning on which we focus our research.

Our approach to the study of learning has been developed within the framework of an understanding of knowledge. From an educational perspective, there are at least two points of justification for such an approach. First, if the learner's conception of a certain phenomenon is at variance with the conception underlying the calculation he or she makes, it severely restricts the learner's understanding of what he or she is doing. Second, because the learner in such a case has simply learned to handle a restricted group of tasks in a certain way, he or she will not be able to use the conceptual tools of scientific thinking, on which the methods for solving the problems are implicitly based (outside the range of those tasks).

In our view, learning (or the kind of learning we are primarily interested in) is a qualitative change in a person's conception of a certain phenomenon or of a certain aspect of reality, it is a distinct change in how that phenomenon is perceived, how it is understood, and what meaning it carries for the learner. However, such fundamental aspects of the subject matter—the very conceptions of the phenomena that statements are made about—are generally not visible in the classroom. They are invisible because they are assumed, and they are assumed because our understanding of the world is a part of us. This is exactly why the students may master certain methods of calculation without having adopted the conceptualization underlying them. In such cases, the students have at least two different conceptions of the same phenomenon—namely, one that is implied by the correct way of handling the calculation and one that is the unchanged common-sense conception that the students brought with them to the study situation. Changes from one conception to another do occur, of course, mainly because the conflict between different ways of thinking becomes explicit due to the problems the students must solve or due to the fact that the teacher deliberately chooses to thematize the conflict. To map the naive (though frequently functional in the context of everyday life) conceptions that can be found among the students and to describe the changes that take place in those naive conceptions is, in our opinion, a most important task for educational research. This research may also increase the likelihood that such changes will occur.

QUALITATIVELY DIFFERENT CONCEPTIONS

In the introduction to this chapter, we said that first we would outline our view of learning and then elaborate its methodological implications. The reality of the research, however, is not a neat sequence from developing an articulated view of the phenomenon to be studied to drawing the methodological conclusions from that view. Rather, by studying the phenomenon, our view of it may change somewhat, which then may lead to some alterations in the research methods adopted, which again may make some new aspects of the object of research visible, which may in turn have a number of implications concerning methodology, and so on.

In our case, a discovery of decisive importance was that for each phenomenon, principle, or aspect of reality, the understanding of which we studied, there seemed to exist a limited number of qualitatively different conceptions of that phenomenon, principle, or aspect of reality. There are always commonalities in the ways in which people who belong to the same culture account for a phenomenon, as well as at least some differences. But on a level of description between the general and the idiosyncratic, there are ways of understanding, that are neither common to everyone nor unique to anyone. What are these qualitatively different conceptions of a phenomena, principles, aspects of reality that exist?

A *conception* is a way of seeing something, a qualitative relationship between an individual and some phenomenon. A conception is not visible but remains tacit, implicit, or assumed, unless it is thematized by reflection. In this sense, conceptions are simply categories of interpretation in terms of which we understand the world around us. In a problem like the one in the previous section, the conceptual aspect is not salient. Because conceptions are categories of interpretation, in order to find out what they mean to each student, we must let the students choose their own interpretations. The fewer conceptual constraints we impose on reasoning as a result of the problem itself, the better. The more natural the questions we ask and the more sensitive our listening to the answers the students give, the greater are our chances of understanding how the students understand.

We now exemplify at some length the characteristics of qualitatively different ways of perceiving and reasoning about certain phenomena, as well as qualitatively different conceptions of certain aspects of reality. One aim of an ongoing investigation is to find out what preconceived ideas students at a technological university hold about the physical reality before taking part in mechanics courses, and how, and to what extent, their notions are modified as a function of the educational experience offered by those courses.[1] In one of the substudies, two sets of interview questions were constructed, with four questions in each set, each question forming a pair with a corresponding question in the other set. The questions within each pair were identical in structure but different in content. Here, we discuss questions in only two of the pairs. In the first case, the students were asked to comment on a physical event with a body moving at a constant velocity: in the second case, there was a body in decelerated motion.

Thirty students were interviewed twice; the first time at the beginning of their studies and the second time after their having completed the first

[1]This investigation, called "Study Skills in Mechanics," is being carried out jointly by staff from the Department of Mechanics, Chalmers Technological University, Gothenburg, and the Department of Education, University of Gothenburg. The project is directed by Svensson, and some of the results reported here have been published previously in a working paper by Johansson (1981), who has been a research associate in the project.

course in mechanics. On each of the two occasions, one of the two alternative questions within each of the four pairs was asked, and the order was reversed for half of the group of students to make the two interviews comparable at the group level. Fifteen students were asked the first question on the first occasion, and 15 the second one. On the second occasion, the questions were reversed for the two groups of students,

1. A car is driven at a high constant speed straight forward on a motor-way. What forces act on the car?
2. A cyclist is cycling straight forward at a high constant speed on a road. What forces act on the bicycle?

Let us now consider the dialogue that took place between the experimenter (E) and one of the subjects (S 2) in connection with the first question:

E: If we take a car being driven at a constant speed straight forward, a high constant speed straight forward on a motorway, what forces act on the car?
S: Motive power from the engine, air resistance, frictional force on all the bearings and so on, gravity, and normal force.
E: How are they related to each other?
S: Gravity and normal force are equal, and the engine is used to counter-balance the sum of the air resistance and the frictional force.
E: Is there anything left over?
S: Well, nothing important.
E: Why is that?
S: *When he drives at a constant speed, all the forces counterbalance each other.*
E: But isn't there any. . . .
S: I suppose there's heat in the air, perhaps.
E: Yes. Isn't anything used for, some force for [moving forward], mustn't one have more, as it were, in order to [move forword], than, as it were, backward if you're going to get the car to move forward at all?
S: Well, yes. When it accelerates, more power is needed forward than when it is moving at a constant speed.

We can now compare this with an excerpt from an interview with one of the other subjects (S3).

E: A car is driven at a high constant speed straight forward on a motorway, can you draw the forces acting on the car?
S: A car.
E: Hm.

S: Viewed from above, then?
E: Hmmm.
S: On a motorway?
E: Hmm. Ye-es.
S: Well, we have gravity drawn straight down there . . .
E: OK.
S: In a point.
E: Hmm.
S: And then there's air resistance, right. . . .
E: Hmm.
S: Then friction against the road surface where there is also some resistance. Then there's. . . .
E: Now let's see, I'll call the air resistance 1 and the friction against the road surface, you write that there yes, an arrow which I shall call 2 . . .
S: I'd draw it like that too.
E: Yes.
S: It'll be the same here against the wheels.
E: All of them are 2, yes?
S: Hmm. Then the car is moved forward by the engine, then.
E: Hmm.
S: *And then a force which is directed forward which has to be greater than those there. Number 3 thus has to be larger than number 1 and number 2, otherwise it wouldn't move forward . . . ,*
E: So that the car's, the force that moves the car forward is larger than those in the wheels as you said and this together. . . .
S: Yes, they have to be.
E: And then you had a force directed downward, that was gravity.
S: Yes.
E: Yes. Have you got anything more?
S: Hm. Well, Then it could be windy, there could a wind too, if you drive across large open fields and the wind's blowing . . .
E: Ye-es, exactly.
S: I suppose it's a good idea to put them all down, in any case.

What is the main difference between the two students' ways of dealing with the problem? Quite obviously, from the point of school physics, the former is right and the latter is wrong. To describe the answers in terms of right and wrong is, however, hardly what we mean by "revealing the students' conceptions."

On one level, the difference between the two students' ways of reasoning is striking. According to the former, the magnitude of force in the direction of the movement equals the magnitude of the forces in the reverse direction, so that the moving body is in equilibrium at a constant velocity. According to the latter, the resultant of the forces acting upon the car must exceed zero and must have the direction of the motion, otherwise the body would not move. In his understanding, a body is in equilibrium only when it is at rest.

We can, however, go further in trying to characterize the qualitatively different ways in which the students apprehend the actual problem. In order to do so, we must take the entire range of interviews into account. Together, they form the whole from which the pattern of qualitatively different modes of thinking can be discerned. We observe that the students who seem to realize that opposite forces are in balance have chosen "velocity" as a point of departure and they have compared constant velocity to acceleration and deceleration. They have explained that the equilibrium of forces makes the velocity remain constant (rather than cause acceleration or deceleration).

The other students, however, seem to have considered the problem from the point of view of "motion" and they have compared motion with rest. They have explained that the greater force forward makes the object move (rather than remain at rest). The most fundamental difference between the two students' understanding of the problem is thus reflected in the difference between the two underlined statements in the two interview excerpts:

S 2: When he drives at a constant speed then all the forces counterbalance each other

and

S 3: And then a force that is directed forward which has to be greater than those there . . . otherwise it wouldn't move forward

> There were thus two qualitatively different conceptions of "a body moving at a constant velocity" found in the investigation. (These applied to both of the two structurally identical problems above: the one with the car and the one with the bicycle.) A body in this kind of motion was apprehended either as (a) **having a constant velocity, due to the equilibrium of forces** or (b) **moving, due to a "motive inequilibrium" of forces.** As we can see, the two conceptions (a) and (b) represent, in actual fact, two kinds of correlated differences, which we may call the "what" and "how" aspect of the event. One regards what is focused on (velocity or motion). The other refers to how the explanation is given; in terms of equilibrium at constant velocity or equilibrium at rest.

TABLE 14.1   Two Conceptions of Movement

	WHAT (is focused on)	
	Velocity	Movement
HOW (explanation is given)		
Equilibrium at a constant velocity	(a)	
Equilibrium at rest		(b)

LEARNING AS A CONCEPTUAL CHANGE

We have introduced a view of learning as a change between qualitatively different conceptions, and we have given an example of learning according to our view: A change from conception (b) to conception (a). The direction of the change is self-evident: learning must, in some comparatively well-defined sense, represent an improvement.

In the aforementioned investigation from which our example is taken, one of the aims was to study in what way and to what extent learning in our sense is taking place as a function of the students' educational experiences. The idea was to compare the students' conceptions of the same physical phenomena on two occasions: before and after a course in mechanics. Differences in conceptions in the right direction between the two occasions would represent cases of learning.

The first interview was in itself a learning opportunity, however. The interviewer's questions made the students reflect, and the questions sometimes made the inconsistencies of their own way of reasoning obvious to them. In this way, it was possible to observe several cases of learning during the interviews, in the sense of a change from one conception of a phenomenon to a better conception of the same phenomenon. (For the moment, we leave the difficult questions of the permanence and generality of such changes. These issues are dealt with in the second part of this chapter).

In the following case, we can see how the shift takes place in the student's (S 11) way of reasoning:

E: So a car is driven straight forward at a high constant speed and you are to draw or tell me what forces act on the car.

S: Well, it's wind resistance, then.

E: Ye-es. What is that?

S: It's the air's particles which . . .

E: Ye-es.

S: Act on the car in some way.

E: Ye-es.

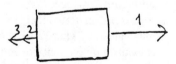

S: And it sort of stands still in relation to the car.
E: What effect does it have, then?
S: Against the direction of the car.
E: I see.
S: And the engine, or the motive force, then.
E: Hmhm.
S: Due to the engine functioning in the car's direction. Then there are different sorts of friction in the car itself which act against, or it can look like it's sideways.
E: Yes. How is the motive power, how great is it?
S: Yes, it's much greater than the others.
E: Ye-es. Individually or together?
S: Together.
E: If you have any more forces here, now you've said—you can write it down, by the way, it would be interesting to see approximately how you. . . .
S: Yes. Then we can say that there's a car here, then. . . .
E: Yes. You're drawing it viewed from above.
S: Exactly.
E: Hmm.
S: Then we can say that Number 1 here is the motive force. . . .
E: Hmhm.
S: Shall we draw them as vectors, then?
E: Yes, it's alright if you draw an arrow.
S: And so that the wind comes from in front.
E: Yes.
S: Then it's wind Number 2.
E: Yes.
S: And there we have Number 3 when there's friction in the bearings and so on.
E: Ye-es.
S: Yes it will, and between the wheels and the road surface.
E: Can it be moved like you did?
S: Yes, one can add those two together.
E: Yes. And how were Numbers 2 and 3 in relation to Number 1?
S: They were clearly smaller.
E: And have you got any more forces that act on the car?
S: We-ell, *I'm not sure about that, you know, maybe they're equally great since the car is moving at a high even speed,* I'm not sure about that.
E: How come you, . . . how did you think when it is equally great . . . when it's greater?
S: Since the car moves forward.

E: Yes.
S: But since it doesn't increase in speed perhaps it—they're equally great instead.
E: Hmm,
S: Probably, Number 1 is greater only when it's accelerating.

Observe in particular this student's answer to the question of how he was thinking when he said that the force forward was greater than the air resistance and the friction:

S: Since the car *moves* forward.

When he then chooses to look at a special type of movement:

S: Since it does not increase in speed [he abandons his original model]. Probably, number one is greater only when it's accelerating.

In spite of the fact that four changes of this kind occurred during the first series of interviews, the participation in the course in mechanics did not affect the frequency distribution across the two categories. Of the 22 students who answered one of the two questions about "bodies moving at a constant velocity," on both occasions, 7 exhibited conception (b) the first time and 6 the second time.

The second pair of questions in the investigation further illustrate conceptual changes as effects of education. We would like to briefly deal with these two alternative cases of "bodies in a decelerated motion." They were the following:

A goods wagon has been derailed and rolls straight forward on a completely horizontal track. What happens?

and

A puck has left an ice hockey stick and glides straight forward on smooth ice. What happens?

We can deal with the qualitatively different conceptions revealed by the two questions without separating them.

There were a number of students whose way of reasoning was in accordance with accepted physical thinking. At least in this case, they seemed to apprehend "bodies in a decelerated motion" as (a) **having a velocity that diminishes due to forces opposite to the direction of motion.** As in the case of conception (a) in the previous example,[2] this way of reasoning takes its point of departure from the "velocity aspect" of the physical event and then only the forces opposite to the motion need to be taken into account.[2] In other cases, however, the students fail to discern velocity as the relevant aspect of the event and focus on the motion in a more global way. Several students reason in terms of a force that the wagon got from the train or the puck got from the stick. This is a force that is inherent, stored in the wagon (puck) and qualitatively different from the external force by which the train affected the wagon (or the stick affected the puck) or by which the wagon (or the puck) may affect other objects. (Nevertheless, it is occasionally used in a sense similar to external forces with which it is compared). In some cases, this internal force is seen as being less than the sum of the opposite forces and it is not seen as a cause of the movement, but rather as an aspect of it. Due to the opposite forces, this force diminishes and the body stops at the same time as the force disappears. In other cases, however, the inherent force is thought to be greater than the opposite forces and is seen as a necessary cause of the movement. The force gradually diminishes due to the opposite forces and when the former equals the latter the body stops. This model of equilibrium at rest clearly resembles conception (b) in the previous example with constant velocity (think of the direction of motion being from left to the right:

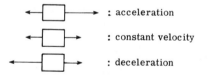

The previous conception seemed, however, to be based on a different model of equilibrium:

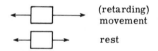

A strikingly Aristotelian-like way of thinking was shown by a student (S 7) who in relation to the "puck question" reasoned in terms of an external force pushing the puck all the time:

E: It's a puck that leaves the ice hockey club and glides straight forward on smooth ice, what happens?

[2]Notice that the labels a, b, etc., do not refer to any general conceptual levels. They are used to denote conceptions specifically related to certain phenomena. This means that two conceptions originating from two different problems but labeled in the same way, as b, for instance, do not necessarily have anything in common with each other, though, it may happen that they have.

S: A puck that leaves . . . .

E: It has left the ice hockey stick and is gliding straight forward.

S: Ye-es, there is a force *that pushes the puck forward.* . . .

E: Yes

S: So then it moves forward. Then there is a frictional force behind but since it's ice I suppose it's not particularly great so then . . .

E: Well. Yes.

S: Gravity and normal forces and . . .

E: Yes.

S: Those sort of things have an effect too.

$$\mu N \longleftarrow\!\!\!\boxed{}\!\!\!\longrightarrow F \qquad \uparrow N \quad \downarrow G$$

E: I see. It's a, yes, perhaps you can draw here. Where do you have the puck here, then? Yes, how does one draw it, like that perhaps.

S: Ah, that wasn't good, this is how I draw it.

E: So it's on its way, then. Can you draw these forces like this?

S: We could call this a normal force, right?

E: Yes.

S: So the gravity in—I'll fix it.

E: Hm.

S: This is a bit stupid, but this here is what we can call it instead, it is the motive force (*F*), then.

E: Hmm.

S: And here behind we have one too, fric . . . what's it called, frictional force?

E: Yes.

S: Moving again, but we can call it G.

E: Hm.

S: It must be those that act on it just then, I think.

E: I see. How are these forces related to each other?

S: They offset each other, the normal force and gravity. . . .

E: Yes.

S: Since they are, . . .

E: Yes.

S: I think, because it's gliding on the ice it has to stay still there on the ice, so to speak.

E: Hmhm.

S: If there isn't a hole in the ice.

E: Hmhm.

S: But this force (μN) since it's ice *must it be very much smaller than this force* (F) *because it's the one that's the greatest and the puck is going in that direction.*

E: Hmhm. This frictional force, where, where does it act on the puck?

S: Underneath, between the surface of the puck and the ice, as it were, that surface under there.

E: Yes. Hm. Well, what happens to the puck after a while?

S: It comes to a halt by itself.

E: Yes. I see. Hmhm. How come?

S: Well. Yes. This . . . ye-es this force must . . . well, this force doesn't increase so it has to be the one that diminishes.

E: Hmhm.

S: I don't really know why it is—but it does stop.

This third force-related conception has the model of equilibrium in common with the second conception. By focusing on the motion (and not on the change in velocity), the student tries to explain it in terms of a force in the direction of motion that exceeds the opposite forces. Equilibrium means rest.

In addition to the correct one, we have thus discerned three other qualitatively different conceptions. Bodies in a decelerated motion can also be interpreted as (b) **having an inherent force that diminishes due to totally greater forces opposite to the direction of motion**, (c) **having an inherent motive force that diminishes due to totally smaller forces opposite to the direction of motion**, or (d) **having an external pushing force that diminishes due to totally smaller forces opposite to the direction of motion.** In other cases, students discuss the problems in terms of kinetic energy. Energy is then dealt with as a directed, vectorial entity and not as a scalar one. There are again two different conceptions, one of which is structurally similar to the preceding conception (a). The difference is that the reasoning is here phrased in terms of energy instead of velocity. The wagon (or the puck) has received a certain amount of energy. The friction and air resistance are working against this forward-directed entity and ultimately make it disappear.

Just as energy was used here in the same way as velocity in the problem solutions based on conception (a), there was one case where, in the student's account, energy had exactly the same functional role as the inherent force in solutions revealing conception (c). According to this way of reasoning, the puck has received energy from the stick. This directed energy is diminishing due to the effect of what is called "friction energy," which is also a directed entity, but opposite to the former. The puck is moving because kinetic energy is greater than friction energy in the beginning. When the two become equal, the puck stops. According to the last two conceptions to be presented here, "bodies in a decelerated motion" can also be interpreted as (e) **having a kinetic energy that diminishes due to forces opposite to the**

TABLE 14.2   Six Conceptions of Movement

	WHAT (is focused on)			
	Velocity	Inherent force	External force	Movement energy
HOW (is explanation given)				
Equilibrium not considered	a			e
Equilibrium at a constant velocity		b		
Equilibrium at rest		c	d	f

direction of motion or (f) **having a motive kinetic energy that diminishes due to totally smaller energies opposite to the direction of motion.** As in the case of constant velocity in the aforementioned car example, we can extract two correlated aspects of the conceptions, previously called the "what" and "how" aspects. The former refers to *what* is focused on: velocity, inherent force, external force, or kinetic energy; the latter concerns *how* the explanation is given: without reasoning in terms of equilibrium, in terms of equilibrium at constant velocity, or in terms of equilibrium at rest (See Table 2). It should be noticed that the different combinations of *what* and *how* dimensions must be interpreted in relation to the problem or phenomenon that the conceptions are applied to. For instance, focusing on velocity and giving an explanation in terms of equilibrium at constant velocity were the most frequent (and correct) ways of handling the preceding car problem, whereas this combination, for obvious reasons, does not occur in relation to the two problems on decelerated motion. We should also bear in mind the fact that the ordering of the conceptions does not necessarily reflect their proximity to the scientifically accepted conception.

As we can see, the pattern that the different conceptions make up together (called *the outcome space* in our terminology) is more complex here than was the case with the problem regarding movement at a constant velocity.

Contrary to the case of the previous problem concerning the students' understanding of bodies in a decelerated motion, the educational effects were most striking. Almost all the heterogeneity in the students' thinking in this respect was abolished by the course. In the interview after the course, all but one of the students took their point of departure from velocity when reasoning about bodies in a decelerated motion and they gave their explanations in terms of forces opposite to the direction of motion only. The answer to the question of why the students' thinking was unaffected in the first case, while it was radically altered in the second has to be sought in the difference in the content of the problems and their relation to the content of the course. To present this analysis here would, however, go far beyond our aim of referring to the actual investigation. Our intention was simply to use it as a concrete exemplar of our view of learning, the methodological implications of which we develop in the next part of this chapter.

METHODOLOGICAL IMPLICATIONS

EXPERIENTIAL PERSPECTIVE

As should be obvious from the preceding examples, when we say that we want to describe the students' ways of thinking or the changes in their ways of thinking, the expression, "describing ways of thinking," is used in a rather specific sense. Namely, we are not trying to describe how the students' thinking in general appears to us (for instance being formal operational, field-dependent or abstract), but we are trying to describe how students think about specific problems and phenomena from their own perspectives. We have discussed two different ways of apprehending a body moving at a constant velocity. Two different ways of thinking refer, in this particular case, to two different interpretations of what is thought about (i.e., a body moving at a constant velocity). This implies that the object of our studies is human subjectivity and we must adopt what has been called a second-order perspective (Marton, 1981). It means that we are not making statements about X (in this case, for instance, bodies moving at a constant velocity) but about people's ideas about X.

Adopting a second-order perspective implies, furthermore, that we are not trying to look into the learner's mind, but we are trying to see what he or she sees; we are not describing minds, but perceptions; we are not describing the learner, but his or her perceptual world.

DESCRIBING THINKING IN TERMS OF ITS CONTENT

The corollary of what has been said in the previous section is that thinking should be described in terms of its content. (*Content* in our sense of the word refers to the thinker's understanding of that which is thought about). In our interpretation, this idea was implicit in the concept of intentionality, introduced by the German philosopher Franz Brentano 1874. This concept was subsequently laid as a groundwork for the phenomenological movement by its founding father, Edmund Husserl. According to Brentano, what distinguishes psychological phenomena from physical is that the former always have a kind of directedness; love is always the love of *someone*, learning is always the learning of *something*.

Using the specific content of thinking in order to characterize thought differs considerably from the commonplace theories in cognitive psychology where, by tradition, the task of research into thinking is to clarify the general mechanisms of thinking and describe the general structures of thought determining and defining the conceptions. Thinking in most cognitive psychology theories is thought of and described by using general mechanisms and as the application of general structures to various content. According to this line of reason, we should be able to find out what it takes to think or learn about different content by finding the mechanisms and structures.

Though there are obviously more or less general aspects of thinking and learning, we should not jump to the conclusion that there are some corresponding general mechanisms or structural entities within the individual. If we consider how individuals deal with various contents in various contexts, we may very well observe that certain aspects crop up repeatedly. We may then conclude that those frequent recurring characteristics tell us something important about human beings' mental equipment. Or, we may conclude that they tell us something important about ways of functioning in relation to various content in various contexts. This latter conclusion is certainly one that we would favor as most warranted.

The difference between the two conclusions—subtle as it may seem—is a very important one. An example sheds light on its nature. We may observe that some children fail to solve a certain problem while other children have no difficulty at all. The children who fail to solve a certain problem (e.g., what makes things sink or float) fail because they are dealing with two different aspects of the problem (e.g., volume and mass) one at a time without being able to bring them into relation with one another. One "explanation" might be that the children who failed are preoperational children. An alternative conclusion to be drawn is that relating two separate aspects of the problem is necessary for its solution and it represents a difficulty to some children in this case. The first interpretation would lead us to expect the children who failed to solve one problem also to fail in solving other problems requiring a concrete operational level of functioning. The second interpretation would make us sensitive to the question of whether the relating of two separate aspects of the same problem is a frequent difficulty in everyday situations (not necessarily for the same children in all situations, but for some children in some situations and for other children in other situations[3]). If this indeed turned out to be the case, we would reasonably make this specific capability an object of instruction.

Apprehended content is the language in terms of which thinking or learning is described. Therefore, when we want to relate our findings concerning

[3]This does not mean that we wish to exclude the possibility that ways of functioning could be generalizable across situations. We only want to make generalizability into an empirical question instead of accepting it as an axiom.

ways of thinking about certain phenomenon, we should not neglect the dimension of content. Instead of trying to either relate the findings to or explain them in terms of underlying mental operations or cognitive structures we would focus our attention on ways of thinking about some other related phenomena or on some other aspects of the same phenomenon.

In the case of our previous physics problem, for instance, it would not add very much to our understanding, at least not from a didactic point of view, if we learned that focusing on velocity was more common among boys, formal operational thinkers, or those coming from better socioeconomic conditions. Nor would it be of much help if it was found that focusing on movement makes less demands on short-term memory capacity.

If we wanted to understand better why focusing on velocity is more difficult than focusing on movement, however, we could reflect along the following lines. *Movement* is of more global character than *velocity*. The latter is in actual fact an aspect of the former and has to be abstracted from it. Furthermore, *velocity* is a derived concept, which is defined as a quotient between two inhomogeneous magnitudes: distance and time. Euclid had expressed the view in Book V that we should only compare homogeneous magnitudes by forming their quotient. John Wallis, in the seventeenth century, was one of the first mathematicians to divide inhomogeneous magnitudes. He may have been the first to write the formula defining velocity (Thompson, 1983).

DISTINGUISHING BETWEEN CATEGORIES OF DESCRIPTION AND WHAT IS DESCRIBED

Conceptions, which make up our unit of analysis, refer to whole qualities of human–world relations. They also refer to the qualitatively different ways in which some phenomenon or some aspect of reality is understood. When trying to characterize these conceptions, we use some categories of description. These categories are, however, not identical with the conceptions— rather, they are used to denote them. (When using expressions such as "ordering the conceptions," we were thus in error. We should have said "ordering the categories of description" instead). To the extent that conceptions reflect the terms in which people interpret the world around them, categories of description express our interpretations of others' interpretations. As our interpretations may be more or less correct, there is a certain relativity linked with the choice of the categories of description: they have a considerable degree of autonomy in relation to what they refer to (i.e., ways of understanding something).

Categories of description are invented to characterize conceptions found in concrete situations, but these categories may be lifted out of the context where they have been found or invented in order to be used as tools for understanding conceptions of similar phenomena or aspects in different situations. The fact that we have to be able to disregard great differences in order to see what is similar should be obvious from our previous comparison of a way of thinking implied by Euclid's mathematics and a way of thinking implied by Swedish students' answers to interview questions in the course of an investigation carried out in the early 1980s.

CATEGORIES OF DESCRIPTION AS RESULTS

As implied by the previous section, we certainly do not consider the construction and derivation of categories of description as a simple reading off of what is in front of our eyes. Quite the contrary: categories of description have to be discovered (in actual fact, it is the conceptions that are discovered) and they make up the main results in the kind of investigations we are discussing here.

According to the canons of research methodology in the behavioral sciences—and also in accordance with common practice—categories of description have to be defined in advance. We must have the categories to start with. What we then do is find relationships among the categories of description (which are usually expected to be of a quantitative nature). If our investigation is derived from theory, then we may test some hypotheses about relationships; if our investigation is exploratory, we must try to find the relationships more or less by trial and error. However, the research situation we have in mind here is a radically different one. Our point of departure is that we want to find out the different ways in which a certain phenomenon or certain aspect of reality is experienced, understood, and conceptualized. From such a perspective, instead of being a ready-made instrument brought to the investigation, categories of descriptions are its main results.

QUALITATIVE VERSUS QUANTITATIVE RESULTS

In an investigation, we may have either qualitative or quantitative results, or both. The reason for our arguing for categories of description being recognized as results is that qualitative components in a research project are mostly regarded as belonging to its exploratory phase. Qualitative descriptions obtained by the interview method or by participant observation are frequently seen as intermediary steps toward obtaining the real (quantitative) results, such as the testing of the hypotheses, the finding of the correlations, and so forth. In our case, the picture is reversed. Both aspects are present; there are the categories of description found (the qualitative outcome) and there are frequency distributions related to those categories (e.g., before and after the course)—the quantitative result. No doubt, it is only the former that is generalizable and is—from the point of view of this chapter—the figure in the figure–ground relation of the two kinds of findings.

RELIABILITY OF DISCOVERY VERSUS RELIABILITY OF IDENTIFICATION

The question is frequently asked: Would another researcher working independently find the same categories and conceptions that we did? However, according to what has been said previously, we consider the finding of a category of description a discovery,[4] and why should we require two researchers to make the same discovery independently? On the other hand, once the discovery has been made, we should certainly be able to communicate it, and other researchers should certainly be able to use the intellectual tools that are supposed to be the outcome of this kind of research and be able to replicate and confirm our discoveries. Consequently, what we want to ascertain is that once categories of description are made explicit, other researchers should be able to identify them when they are applicable in varying contexts. In accordance with this, indicators of reliability should not concern the extent to which categories are discovered independently, but the extent to which they are identified once they have been specified.

There are (at least) two problems involved in giving indicators of interjudge reliability in terms of percentage agreement. First, as it was obvious from our excerpt from the interview with (S 11), there may be more than one category applicable to an answer or to a protocol. Second, in other cases, the answer to a certain question or the entire protocol may not contain sufficient information on which to base a judgment concerning the applicability of any category. Furthermore, we should observe that a percentage agreement due to chance alone varies as a function of the number of the categories. (It is 50% for two categories, 33% for three, 25% for four, etc.) As a very general estimate, we could say that, to be satisfactory, reliability figures of this kind should be within the range 75–100%. The results obtained in our research group are within this range.

POINTING OUT THE NATURE OF OUTCOMES VERSUS DESCRIBING METHODS FOR ARRIVING AT THEM

Another implication of the view that categories of description are discovered is that we cannot specify methods that would ensure a certain

---

[4]In actual fact, it is conceptions and not categories of description that are discovered. From a stylistic point of view, it is, however, somewhat awkward to maintain this distinction consistently.

predetermined outcome. A discovery can never originate from an algorithmic process. Our aim is not to measure certain qualities of individuals but to reveal as yet unknown qualities of their relations to the world. There is simply no way of giving highly specific instructions for how such an enterprise should be carried out. There is, however, an example of very thorough and detailed analysis, carried out by Theman (1983), illustrating the considerations taken into account in the search for categories of description characterizing different conceptions of political power. Some details of that research work were recently described by Svensson and Theman (1983).

In our research group, we have used mostly interviews, but there is nothing essential about this specific way of collecting data. What is fundamental, however, as was suggested earlier, is that the procedure used should be sufficiently open to allow the subjects to express their own ways of structuring the aspects of reality that they are relating to and to give them the opportunity to choose the terms in which they interpret the situation they are facing.

The questions asked are of decisive importance, of course. Surely, it takes bright ideas and creative insights to formulate questions that stimulate openness. Also, carrying out the kind of research interview we have in mind is a delicate task. The fact that we are interested in *what* the students think about and *how* they think, not in whether or not they manage to produce the right answer should be obvious to them as well. The interview should be an open-minded exploration of the landscape of thoughts, not an examination nor an instructional session.

All the conversation is subsequently transcribed and the transcripts are analyzed. The protocols together form the whole; they give the context and background against which single statements gain their significance. The aim is not to give an exhaustive description of each interview, but to find especially relevant points. By reading through the protocols several times, a selection procedure is employed. We try to pick the statements that are relevant in relation to the question, "What are the different ways in which the actual phenomenon or aspect of reality is understood?" The interpretation of each statement has to be made in relation to its context in the protocol where it appears. As we said earlier, the statements gain their significance through comparisons with other statements in the same protocol and in others that were found to be relevant. As a result of such comparisons made on the basis of the interpretative work, a pattern of similarities and differences may emerge. Once such a pattern has appeared, however vague its form may be, it is used as a conjecture for a systematic search for, and control of particular conceptions in each protocol.

In this chapter, we offer general methodical recommendations and try to characterize the general methodological orientation. Above all, however, we point to the kind of outcomes we are seeking.

DISTINGUISHING BETWEEN LOGICAL AND EMPIRICAL RELATIONS

From the preceding distinction between the categories of description and the object of that description, another distinction follows—namely, that between logical and empirical relations. The former is to be seen between the categories themselves, the latter between the instances referred to by the categories. For example, one of the ways of apprehending bodies moving at a constant velocity was found earlier to be characterized by a person focusing on the velocity aspect of a physical event. Also, focusing on the velocity aspect was found to be a distinctive feature of one of the categories. This concerns the logical relation between the two categories. To what extent persons who focus on the velocity aspect in one case also do so in the other case is, however, an empirical question.

There are two usually unstated presuppositions that make it questionable as to whether we can expect the logical relations to hold also on the empirical level. Such an expectation would be based on two hidden assumptions. First, we must assume stability in the individual—that is, to assume that he or she thinks in the same way about the same thing at two different points of time. Second, we must assume that the logical definition of the problem is the only relevant point of view for the individual. If we consider the second point first, we have to conclude that the assumption is frequently (not to say mostly) unwarranted.

In the investigation to which we have been referring, the second problem was aimed at revealing the students' understanding of bodies moving at a constant velocity. A bicycle was the object in this problem. Though the car problem and the bicycle problem can be considered identical from the point of view of physics, this second problem turned out to be clearly more difficult in the sense that more students gave answers in terms of equilibrium at rest (i.e., answers reflecting conception [b]), both before and after the course. In actual fact, some students who had given answers to the car problem in terms of equilibrium at a constant velocity (i.e., answers reflecting conception [a]) before the course, turned to the Aristotelian-like variety of solution when confronted with the bicycle problem after it. Our interpretation is that these differences have to do with the students' everyday experience of cycling, The tendency to assume the need for a greater force forward probably increases if one has experienced the situation firsthand where one has to provide a force in order to maintain a forward movement.

Teaching in school is obviously aimed at enabling the students to disregard, in various situations, what is not relevant from the perspective of a discipline-bound definition of those situations. Though clearly within the framework of a highly specific context—the context of the school—they are expected to learn to function in an acontextual manner.

We have been able to avoid questions concerning the endurance and generality of changes by saying that we are only interested in describing relations between the individual and certain phenomena or aspects of the world. In order to be able to claim that learning has taken place (which was at least partially our interpretation of the before- and after-course differences) we must however, make some assumptions about the changes having some permanence and generalizability.

Without abandoning our relational view we could concur in the view of learning as being a reasonably durable and generalizable change. This change does not, however, have to take place *in* the individual but rather *between* the individual and the world. What is necessary is to establish relationships on such a fundamental level that they remain unaffected by the ongoing flux of the situational variations. Such changes can be visible only if we do not merely look at the individual, but rather try to see his or her world.

We would like to argue that the distinction suggested by the title of the present section leads to a fairly radical stance from a methodological perspective. Usually, empirical relations are hypothesized on general psychological grounds or they are sought for by trial and error in exploratory studies. In our case, the starting point is in the relationships of a logical or internal character, and we must explain especially the cases in which the relationships hypothesized do not hold on an empirical level or when such relationships are not found in exploratory studies. What is taken for granted and what has to be explained shifts thus in a figure–ground fashion. This is illustrated by the case of rest and motion (i.e., whether rest is taken for granted and motion is explained or whether motion at a constant velocity is taken for granted and changes in velocity are explained).

Toward a Framework for Describing Learning as Change between Qualitatively Different Conceptions

Results of the kind we have been arguing for in this chapter have indeed been produced in widely varying disciplines. Qualitatively different ways in which people experience and conceptualize various aspects of their reality and various phenomena in it have been mapped by researchers in developmental psychology, anthropology, clinical psychology, history of science, sociology of knowledge, and so forth. The point has been made that we should try to integrate such findings on the basis of content, and by doing so we should arrive at an alternative organization of this type of knowledge. Phenomenography has been suggested as a name for this domain, or family of content-centered domains. (Marton, 1981). This means, for instance, that we should bring together knowledge about the different ways in which people may interpret the physical reality, regardless of whether such differences originate from differences between cultures, historical periods, or age levels. By doing so we would arrive at something we may call "a phenomenography of physics";[5] it would contain layers of meaning linked to physical theory, a description of objects that are denizens of Popper's "third world."[6] Such a science of categories of description (to which investigations of the kind illustrated in this chapter contribute) should be of considerable didactic interest, at least if one adopts the view of learning that we have tried to make explicit in our presentation. The assumption is that if we want to make the students think in a certain way about something, it should be useful to know what other ways there are to think about it.

According to the argument for phenomenography, we should bring together findings arrived at by highly differing methods. Phenomenography is thus more or less neutral from the point of view of specific research methods. It represents, however, a distinctive perspective in which findings may be reinterpreted in ways that do not accord with their original meaning. As was shown in one of our earlier examples, a way of dealing with a problem (which had been regarded as symptomatic of an underlying cognitive structure) can very well be seen as an aspect of a human–world relation. The fact that phenomenography is basically not method-oriented makes it point away from itself (i.e., from questions of what it is and how it should be done to the kind of findings it is about). It makes it point to a certain type of knowledge, or a certain type of result—knowledge that we have advocated, and results that we have exemplified.

Phenomenography (i.e., various domains of categories of description denoting conceptions of certain kinds of phenomena) should offer an appropriate framework for the systematic description of learning suggested in this chapter. A change between two qualitatively different conceptions could be understood in relation to the background of alternative possible changes

---

[5]In a similar way, we could think of a phenomenography of mathematics or phenomenography of biology corresponding to the systematization of knowledge about the different ways in which people use mathematical or premathematical concepts or conceptualize biological phenomena.
[6]In Popper's (1972) words, we can speak of a world of physical objects and physical states ("the first world"), a world of states of consciousness ("the second world"), and a world of objective contents of thought ("the third world"), which includes the products of the human mind, such as scientific theories, scientific problems, social institutions, and works of art.

represented by a structured set of categories of description. The study of learning, on the other hand, would provide descriptions of empirical relations between conceptions. This would illuminate important variations in meaning between logically similar conceptions and it would help focus on meanings and relations of special didactic relevance.

What is suggested may be briefly illustrated by reference to our example. As was pointed out previously, there are some apparent general logical similarities between the conceptions of the two kinds of events referred to earlier. This is true, above all, as the right conceptions are concerned. The nature of the wrong answers reveals, however, that change from wrong to right conceptions have different meanings in the two cases. This also illuminates differences between the right answers to the two problems—differences that are highly relevant form a particular perspective of learning.

In the case of bodies moving at a constant velocity, the change between the two occasions was reasonable to expect. However, it did not occur in actual fact and was to take place from the students' seeing constant velocity as a case of motion related to a motive force, to seeing constant velocity as a special case of effect related to an equilibrium of forces. This is a change similar to the historical transition form an Aristotelian to a Newtonian view of motion.

In the case of bodies in a decelerated motion, the change between the two occasions, indeed came about. It took place mainly from the students' method of including mass together with velocity in using concepts close to innate force, momentum and kinetic energy, to not including such concepts (and mass), but using a model with only velocity and external force. This change is mainly one of learning which model and which concepts have to be used, but it does not necessarily mean any fundamental change in the conception of motion has taken place.

From a perspective of learning, the relations between conceptions are different in the two cases and therefore also the meanings of the conceptions are different. Not only does phenomenography (i.e., domains of conceptions) offer us a framework for describing learning as change between qualitatively different conceptions, but also the description of such changes enhances our understanding of the nature of the conceptions and the structure of the domains, and thus contributes to an improved basis for didactic efforts.

ACKNOWLEDGMENTS

The research reported here has been financially supported by grants from the Swedish National Board of Higher Education and from the Swedish Council for Research in the Humanities and Social Sciences.

REFERENCES

Johansson, B. (1981). Krafter vid rörelse. Teknologers uppfattningar av några grundläggande fenomen inom mekaniken [Forces at motion. Technological students' conceptions of some basic phenomena in Mechanics]. *Pedagogiska institutionen, Göteborgs universitet, 14.*

Marton, F. (1981). Phenomenography—describing conceptions of the world around us. *Instructional Science, 10,* 177–200.

Popper, K. R. (1972). *Objective knowledge: An evolutionary approach.* Oxford: Oxford University Press.

Svensson, L., & Theman, J. (1983). The relations between categories of description and an interview protocol in a case of phenomenographic research. *Department of Education, University of Göteborg, 02.*

Theman, J. (1983). Uppfattningar av politisk makt [Conceptions of political power]. Unpublished manuscript.

Thompson, J. (1983). Talbegreppets historia från pedagogisk synpunkt [The history of the concept of number from a pedagogical point of view]. Unpublished manuscript.

LEARNING SCIENCE BY GENERATING NEW CONCEPTIONS FROM OLD IDEAS

M. C. Wittrock

INTRODUCTION

Since the mid-1970s, research in science education and teaching has developed an understanding of science learning that differs considerably from earlier, commonly held views. As part of the cognitive movement in instruction and learning (Wittrock, 1978), the research in science education has developed an understanding of science learning that emphasizes the interaction between students' knowledge and thought processes, on the one hand, and scientists' conceptions of physical and biological phenomena, on the other hand.

The recent cognitive approaches to learning and teaching, such as the one that I first wrote a decade ago (Wittrock, 1974b) and since then have elaborated in mathematics learning (Wittrock, 1974a), reading comprehension (Wittrock, 1981) and, with Roger Osborne, in science learning (Osborne & Wittrock, 1983), have tried to conceptualize the nature of this interaction between students' thinking and scientists' data and models. To describe this interaction productively for science education and teaching, many of these recent cognitive approaches to instruction emphasize the importance of understanding that science learning begins with the learners' science. I maintain that, in addition, we must understand how students use their previously learned, often naturally and informally acquired, conceptions of science and ways of thinking to generate meaning for events that scientists explain in alternative and more sophisticated ways. When students' information and ways of thinking are appropriate for learning science in school, then teaching can focus on assimilating new scientific ideas to old cognitive structures. However, when students' knowledge and thought processes are inappropriate and therefore counterproductive for learning science in school, then accommodative learning, in which students learn new scientific conceptions that revise or supplant less adequate, previously acquired frameworks, becomes necessary.

The study of these generative processes (assimilative and accommodative) in science learning leads to new developments in (1) the measurement of student and teacher thought processes and cognitive structures, (2) the conceptualizations of science learning, and of what it means to learn science in a meaningful way, (3) the number and nature of the culturally induced or idiosyncratically created alternative frameworks students acquire about science, and (4) the characteristics of teaching and instruction effective for facilitating student generation of scientific conceptualizations from informal viewpoints about physical and biological science.

A REVIEW OF PART II

The six chapters of Part II of this volume treat all four of the areas mentioned in the previous paragraph, with the developments about measurement of thought processes in science treated especially in Part I. In the set of chapters in Part II we learn about (1) models of meaningful science learning as a change in learners' conceptualizations, (2) background ancillary student knowledge needed to make sense of scientific concepts and principles, including cross-cultural differences in student background knowledge about scientific concepts, and (3) the teaching of science concepts as accommodative changes in cognitive structures, as the learning of qualitatively different science concepts, and as the modification of metacognitive thought processes involved in learning how to learn science.

To present the related contributions of these six chapters to the emerging cognitive view of science learning in a unified context, I begin with the view of learning and teaching that these chapters suggest to me. I list and discuss main points that synthesize many of the major findings and models described in the chapters.

The overarching conception of science learning and teaching that I acquire from these chapters is that it is the process of stimulating learners to construct new and better understandings of scientific phenomena by assimilating new concepts to old frameworks (a type of evolutionary learning) or by accommodating new frameworks from old ones (a type of revolutionary learning). With each of these types of learning, teaching is no longer only the presentation of the scientists' view of reality. Neither is teaching the presentation of the difference between the scientists' and the learners' views of reality. Instead, the emerging conception emphasizes the distinctive responsibilities of learners, who construct new meanings from old models and experiences, and of teachers, who know and understand the learners' thought processes and conceptions of science and who provide appropriate instruction to lead the students to assimilate new concepts or to accommodate new frameworks.

The main points that these six chapters contribute to this overarching conception of science learning are the following:

1. We are developing an understanding of science learning that distinguishes it from learning in some other areas of study and that emphasizes the importance of accommodative learning, as well as assimilative learning.

Strike and Posner (Chapter 13) discuss how science learning often involves learning a framework that is different from the one the student has previously learned. As they indicate, that type of learning is different from learning to incorporate information into schemata or into slots. It is learning the relations of concepts to a new theory or new basis for viewing science.

Sometimes, students may not have an appropriate preconception to enable them to understand the new information. As is often the case, they do have a preconception that is inappropriate and counterproductive. The example Strike and Posner gave of Freud's attempts to teach people that mind has unconscious components illustrates these points quite well. Because of Descartes' widely accepted notion of conscious mind, people often possessed a preconception that made it difficult for them to understand Freud's notions about mind.

Freud used hypnosis as a technique to demonstrate that people had thoughts of which they were not conscious and that these thoughts could influence their behavior as evidenced by purposeful mistakes and motivated forgetting. By means of these demonstrations, he hoped to lead people, through accommodative learning, to an understanding of his theory of mind and its implications for viewing mental illness, art, and religion quite differently from earlier notions about them.

Johansson, Marton, and Svensson (Chapter 14) discuss somewhat related issues when they mention that students who organize their thinking around the concept of velocity, rather than motion, approach problems in mechanics qualitatively differently from each other.

Novak (Chapter 12) discusses how assimilative and accommodative learning become meaningful. Using Ausubel's model of meaningful learning, Novak discusses how science learning involves learning relationships among and between concepts and experience. Typical measures of student achievement, he writes, do not usually measure the higher-order relationships that distinguish meaningful science learning from arbitrary and verbatim learning. He maintains, as I do in my model of generative learning, that meaningful learning is a student generative process that entails construction of relations, either assimilative or accommodative, among experience, concepts, and higher-order principles and frameworks. It is the construction of these relations between and within concepts that produces meaningful learning.

Learning new frameworks, making qualitative changes in conceptions, and engaging in accommodative learning occur also in disciplines other than the physical and biological sciences. However, these types of learning are critical in science education, difficult to understand and to facilitate, and involved in the issue of science learning as both as an evolutionary and as a revolutionary process.

2.   Students have culturally transmitted and idiosyncratically generated models or preconceptions of science that influence science learning and teaching.

We should not be surprised when we learn that students believe that "heat molecules flow" and that "a vacuum sucks." These common sense notions explain, at an intuitive level, a wide variety of everyday experiences. As Hewson (Chapter 10) alludes, Lavoisier gave an explanation of heat not far from the notion that heat molecules flow. In some cases, scientists have labored for many years to convince themselves that a non-intuitive theory or model described scientific phenomena better than does a popular or intuitive theory. The theory that invisible microbes pervaded our surroundings, caused disease, and could infect patients during surgery was an elegant, simple, and parsimonious theory that some scientists and surgeons thought was an incredible, ludicrous idea when it was introduced. How equally strange and inscrutable must some modern-day scientific theories seem to students who have constructed alternative intuitively appealing frameworks that they find adequate for explaining and coping with everyday problems.

Hewson provides data to show that cultures transmit different conceptions of heat. The caloric view of heat flow is relatively common among children in the United States, Canada, France, and England. In contrast, a scientifically more advanced conception of heat than the caloric view, a kinetic or at least a prekinetic view, is relatively common among Sotho children in Africa, who have not been taught a kinetic theory of heat.

Roger Osborne's work in children's science, discussed in Chapter 2 (Gilbert, Watts, & Osborne) of this volume, shows some of the student-generated models that exist in physics. To explain how current flows in a simple DC circuit, he finds that young children in New Zealand, England, and the United States characteristically use three or four models, only one of which is like the physicists' explanation that the current flows equally, and in one direction only, throughout the circuit. Osborne finds these models difficult to unlearn. Students vigorously defend them, even in light of demonstrations that the models are inaccurate. Again, accommodative learning of new conceptions emerges as an important area of study in science education.

To facilitate accommodative learning, we need to identify and to measure the number and variety of students' preconceptions of science; and we need to devise ways to use this knowledge about students' preconceptions to teach them alternative, more useful conceptions.

3. The ancillary and tacit knowledge involved in learning concepts and principles in science is being discovered and articulated.

In greater detail than I can properly summarize here, Reif (Chapter 9) explains an ancillary knowledge necessary to make sense of commonly taught concepts, such as acceleration. Without access to the undiscussed and unreported prologues and assumptions known by the instructor, beginning science students delve into physics problems armed mainly with formal statements of equations. As a result they often make mistakes. When students follow the procedure for specifying a concept, such as acceleration, they often better understand the problems and the equation, probably because following the procedure requires them to explain the meaning of the concept.

From a different but related approach, Reif indicates more than a need for meticulous attention to detail in learning concepts and principles in science. He shows how student specification of everyday concepts, such as color, by use of prototypical examples, is poor preparation for the detailed and unambiguous specification of concepts in science, using explicit rules. Again, the students' knowledge and ways of thinking influence science teaching and take it well beyond the simple notion of presenting the subject matter in clear language.

4. The interaction between students' alternative structures and scientific concepts influences the understandings and meaning acquired in science classes.

This variation on the theme discussed in each of these three main points mentioned so far deserves attention because it emphasizes the importance of getting students to relate and to change old conceptions into new frameworks. As Osborne and his associates have found, students often do not revise everyday concepts because they isolate them from the principles presented in the classroom. They feel, for example, that electrical current may flow as the teacher says it does in the classroom. But in the student's home, electical current flows differently, that is, as the student's model indicates that it does. In this way, student preconceptions and teachers' models of electricity do not interact with each other. As a consequence, student frameworks remain unchallenged, cognitive dissonace is not induced, and accommodative learning does not occur, even though a written examination over current flow might indicate otherwise. That is, the students learn how the teacher wants them to behave on the examination, but they do not incorporate or use scientific understandings in their lives outside the school, which is perceived as an artificial place having little to do with the real world.

5. We are beginning to develop new ways to facilitate science learning using cognitive models of teaching and instruction.

From the models and assumptions underlying the research discussed in Part II of this volume, we see the beginnings of highly promising research-based approaches to modifying instruction in science. Novak discusses the effects of concept mapping and Vee mapping strategies upon learning science concepts meaningfully and upon helping students to "learn how to learn," or to use metacognitive strategies.

Gunstone, Champagne, and Klopfer (Chapter 11) report an extensive research program on children in Pittsburgh and adults in Victoria learning to change real world ideas and cognitive structures relevant to solving problems in mechanics. The adults, science majors, changed their cognitive structures more from the pretest to the posttest as a result of instruction than did the children. I am delighted that accommodative learning occurred with the adults and not surprised that it did not occur in this relatively brief period of instruction with the children. The new frameworks were perhaps too different and difficult for the children to learn quickly and meaningfully.

In my studies with teaching kinetic molecular theory to primary school students (Wittrock, 1963), I found that the children learned and transferred concepts about heat and about states of matter to new situations. They also retained the concepts one year later. We studied their beliefs about changes in states of matter and then taught them an organized conception, kinetic molecular theory, associated to numerous real-world examples, using simple analogies and special illustrations prepared by artists. We had the children generate explanations for changes in states of matter, relating the theory to concrete and familiar examples and problems.

In another study (Wittrock, 1967), we taught elementary school students a heuristic to use to solve simple concept identification problems that involved classifying figures. The simple metacognitive strategy they learned was first to generate all possible four answers and then to test each one in turn, eliminating the answers that were shown to be wrong. Compared with other groups taught less effective strategies, the experimental group achieved the greatest amount of learning and transfer.

In the studies by Novak and by Champagne, Gunstone, and Klopfer (Chapter 11), as well as in the other studies I have mentioned in the preceding paragraphs, there are some suggestions about methods to facilitate learning of new concepts and new conceptions in science. First, cognitive dissonance, described by Champagne et al. holds promise. It offers a way to get students' preconceptions to interact with science concepts. However, it is a complicated technique. As Osborne and his associates found, students easily avoid the dissonance by thinking that what they see and learn in the classroom and laboratory does not relate to the same phenomena, such as electricity, in their homes or other out-of-school settings.

A variety of other promising new procedures is also developing, such as Novak's concept mapping. One of its advantages is that it requires students to generate the conceptual relations among concepts and between concepts and examples. Another promising technique is the teaching of metacognitive strategies for solving problems. Yet another technique is the teaching of pervasive, organized scientific systems, such as kinetic molecular theory, to very young children before they have evolved in detail an alternative counterproductive theory. In all of these teaching approaches, it is the interaction between student conceptions and scientists models that is critical to facilitating generative learning.

SUMMARY AND CONCLUSION

In summary, from the studies and arguments presented in Part II, and from my own research, I do not believe that science teaching proceeds as many people formerly assumed that it did. I believe that it is no longer feasible to teach only the scientists' view of reality, that is, to teach the subject matter. Nor is it adequate to teach the difference between the scientists' view and the students' view. Neither of these approaches necessarily involves students' conceptions in interaction with scientists' models.

The more appropriate teaching techniques involve students in the generation of relations between what they know and what they are taught. Teachers do not literally *teach* in the usual sense of the word. Instead they facilitate learners' generation of meaning and understanding by helping them to relate new and old information and conceptualizations to each other. Even information and concepts which are directly taught to students must still be generated by them for meaningful learning to occur.

The editors of this volume carefully selected topics in Part II to discuss these fundamental issues in science learning. The chapters in Part II indicate that in science learning we are beginning to understand what students know about science, how they think about science, and how we can facilitate science learning. The authors of these chapters are asking some of the right questions; and they are getting some useful answers to these questions.

Research in science learning reported in these chapters is going in productive directions.

REFERENCES

Osborne, R. J., & Wittrock, M. C. (1983). Learning science: A generative process. *Science Education, 67*, 489–508.

Wittrock, M. C. (1963). Response mode in the programming of kinetic molecular theory concepts. *Journal of Educational Psychology, 54*, 89–93.

Wittrock, M. C. (1967). Replacement and nonreplacement strategies in children's problem
solving. *Journal of Educational Psychology, 58*, 69–74.

Wittrock, M. C. (1974a). A generative model of mathematics learning. *Journal for Research in
Mathematics Education, 5*, 181–196.

Wittrock, M. C. (1974b). Learning as a generative process. *Educational Psychologist, 11*, 87–
95.

Wtitrock, M. C. (1978). The cognitive movement in instruction. *Educational Psychologist, 13*,
15–30.

Wittrock, M. C. (1981). Reading comprehension. In F. J. Pirozzolo & M. C. Wittrock (Eds.),
Neuropsychological and cognitive processes in reading. New York: Academic Press.

AUTHOR INDEX

Numbers in italics refer to the pages on which the complete references are cited.

A

Adelson, J., 97, *100*
Adorno, T. W., 95, *100*
Albert, E., 159, *160*
Anderson, C. W., 212, *230*
Anderson, J. H., 51, *59*, 66, 88, *88*, 163, *186*
Anderson, J. R., 52, 54, *58*, 63, *88*, 125, *127*, *129*
Anderson, R. C., 66, 73, *88*, 123, *129*, 186, *186*
Andre, T., 53, *58*
Archenhold, W. F., 25, *26*
Arons, A. B., 61, *88*
Atkin, J. A., 200, *207*
Atkinson, E. P., 82, *88*
Ausubel, D. P., 11, *27*, 33, 52, *58*, 90, 95, *100*, 128, *129*, 190, 191, 192, 194, 195, 197, *207*, 259
Ayer, A. J., 222, 223, *230*

B

Bacharach, V. R., 53, *58*
Bacon, Sir Francis, 195, *207*
Baggett, P., 53, *59*
Bates, G., 84, *88*
Benassi, V. A., 64, *90*
Bloom, B. S., 120, *129*
Bogden, C. A., 201, *207*
Bower, G. H., 52, 54, *58*
Brackett, G. C., 150, *150*
Bransford, J. D., 66, *88*, 220, *230*
Brink, B. du P., 158, *160*
Brown, H. I., 195, *207*, 212, *230*
Buchweitz, B., 206, *207*

C

Cahn, A. D., 64, *88*, *89*, 163, *186*
Caramazza, A., 148, *150*

Carr, T. H., 53, *58*
Champagne, A. B., 3, 6, 7, 44, *48*, 51, *59*, 64, 66, 69, 82, 87–88, 88, *88*, *89*, 117, 119, 121, 126, 128, 155, *160*, 163, 165, 167, 178, 182, *186*, *187*, 198, 262
Chi, M. T. H., 68, 72, 88, *89*
Chomsky, N., 107, *115*
Clement, J., 133, *150*
Collins, A., 81, 82, *89*
Cordemone, P. F., *205*
Cullen, J., 199, 200, *208*

D

de Saussure, F., 105, *115*
Deese, J., 51, *59*, 167, *187*
DeSena, A. T., 44, *48*, 167, *186*
Desina, A. T., 44, *48*
Dijksterhuis, E. J., 62, *89*
diSessa, A. A., 133, *150*
Douvan, E., 97, *100*
Driver, R., 3, 7, 11, *27*, 64, 66, *89*, 122, *129*, 154, 155, *160*, 211
Duran, R. P., 85, *89*

E

Easley, J. A., 11, *27*, 64, 66, *89*, 154, *160*, 163, 170, 174, *187*
Eaton, B., 61, *89*
Engelhart, M. D., 120, *129*
Erickson, G., 3, 7, 122, *129*, 154, 155, 157, 158, 159, *160*
Ericsson, K. A., 88, *89*, 121
Estes, W. K., 197, *208*

F

Feltovich, P. J., 68, 72, 88, *89*
Fensham, P. J., 3, 26, 27, 29, *49*, 115
Fillenbaum, S., 166, *187*
Frege, G., 112, *115*

267

McCarrell, N. S., 66, *88*
McCloskey, M., 148, *150*
McDermott, L. C., 133, 148, *151*, 229, *230*
McIntire, W. R., 61, *89*
Mehner, D. S., 53, *58*
Melby-Robb, S. J., 206, *208*
Miller, G. A., 166, *187*
Minsky, M., 104, *116*
Mollura, M. F., 206, *208*
Moreira, M., 198, 200, *208*
Murray, H. T., Jr., 198, *209*

N

Nagel, E., 85, *89*
Neisser, U., 85, *89*, 220, *230*
Neser, G. O., 157, *160*
Norman, D. A., 33, 34, 35, *48*, 66, *89*, 126, *129*, 153, *160*
Novak, J. D., 4, 6, 33, *48*, 53, 59, 114, *116*, 190, 192, 194, 195, 197, 198, 204, 206, 207, *207*, *208*, *209*, 261, 264–265
Novick, S., 155, *160*
Nussbaum, J., 155, *160*

O

Osborne, R. J., 3, 12, 26, *27*, 36, *48*, 51, 59, 88, *89*, *90*, 117, 124, 158, *161*, 259, 262–265

P

Paivio, A., 35, *48*, 53, 59
Peterson, K. D., 61, *90*
Petrie, H. G., 153, *161*
Phillips, D. C., 119, *129*, 195, *209*
Piaget, J., 103, *116*, 120, *129*, 154, 159, *161*
Pienaar, H. N., 157, *160*
Pines, A. L., 4, 5, 6, 17, *27*, 51, 52, 59, 96, *100*, 114, *116*, 117, 118, 121, 198, *209*
Plato, 213, *230*
Popper, K. R., 31, *48*, 195, *209*, 255, *257*
Posner, G. J., 6, 96, *100*, 114, *116*, 195, 198, *209*, 215, 216, 219, 222, 229, 230, *230*, 261
Postlethwait, S. N., 198, *209*
Preece, P. F. W., 167, *187*, 199, *209*
Primrose, R., 61, *89*
Pylshyn, Z. W., 53, 59

Q

Quillian, M. R., 31, *48*
Quine, W. V. O., 13, *27*, 224, 225, *230*

R

Rapoport, A., 166, *187*
Redmore, C., 100
Reif, F., 4, 7, 134, 136, 150, *150*, *151*, 263
Renner, J. W., 61, *90*
Resnick, R., 198, *208*
Reynolds, R. E., 66, *88*, 123, *129*
Ridley, D. R., 190, 194, *209*
Riegel, K. F., 186, *187*
Rokeach, M., 95, *100*
Rosch, E., 109, *116*
Rowell, R. M., 33, *48*, 198, *209*
Rumelhart, D. E., 34, 126, *129*
Ryle, G., 53, 59, 119, *129*

S

Sanford, R. N., 95, *100*
Schallert, D. L., 66, *88*, 123, *129*
Schapera, I., 157, *161*
Scheffler, I., 85, *90*
Schwab, J., 206, *209*
Shavelson, R. J., 51, 53, *59*, 119, 125, *129*, 154, *161*, 166, 167, *187*
Shayer, M., 96, 100, 159, *161*
Shuell, T. J., 121, 128, *129*
Siegler, R. S., 123, *129*
Simon, D. P., 229, *230*
Simon, H. A., 88, *89*, 121, 229, *230*
Singer, B., 64, *90*
Smith, E. L., 212, *230*
Sola, J., 53, *58*
Solomon, C. A., 64, 88, *89*, 163, *186*
Speller, K. R., 122, *129*
Squires, D. A., 44, *48*, 167, *186*
Stanton, G. C., 166, *187*
Stevens, A. L., 81, 82, *89*
Strike, K. A., 6, 96, *100*, 195, *209*, 212, 213, 215, 216, 219, 229, 230, *230*
Sutherland, J., 61, *90*
Sutton, C. R., 4, 6, 30, *48*, 99, *100*, 117, 121, 122, *129*, 156, *161*
Svensson, L., 6, 14, 236, 252, *257*, 261
Symington, D., 206, *209*

SUBJECT
INDEX

A

Abstract thought, 110
Accommodation, 215, 221, 228–229, 259–261
Accretion, 126
Affective, 96, 117
Alternative conceptions, 3, 86, 111, 154–155, 163, 184
Alternative frameworks, *see* Alternative conceptions
Algorithms, 30, 32, 37, 38, 39
Ambiguity, 13, 14
Ancillary knowledge strategies, 5, 133–135, 147, 260, 263
Analogues, 85, 86, 155, 216, 219
Anomolies, 216, 219
Apprehended content, 248
Assimilation, 215, 248, 259–261
Associations, 56, 170, 173

B

Behavioral theories of learning, 103, 191
Belief systems, 94

C

Categories of description, 249
Cognition, 101
Cognitive
  dissonance, 263
  elements, 52
  psychology, 51
  objectives, 69
Cognitive structure, 29, 33–38, 51, 86, 96, 99, 101, 113, 172, 173, 191, 259
  accord with reality, 54
  change, 163, 164, 177, 178

components, 33
dimensions, 45, 47, 51, 54–57, 124
descriptive techniques, 29–44, 117, 166–172
dynamic model, 53
eliciting, 38, 43
experts, 37
extent, 54
idiosyncratic, 36
internal consistency, 54
pre-instructional, 174
probes, 165–172, 169, 175
quantitative indices, 45
representation, 38, 39, 40–43, 45, 178–179, 182
Commitment, 95
Computer metaphor, 31
Concept, 36, 38, 53, 63, 67, 101, 108–111, 133, 165, 166
  concrete, 108
  existing, 216
  generative, 259
  labels, 36, 38, 191
  maps, 5, 37, 174, 198, 201, 203, 206, 264
  naive, 235
  organization, 212
  prior, 96, 215
  value, 139–140
Concept structuring analyses technique, 167, 171, 174, 175
Conception, *see* Concept
Conceptual
  change, 2, 110, 183, 185, 210–217, 222, 225
  ecology, 154, 216, 228
  framework, 110, 111, 199, 206
  learning, 3
  learning skills, 150

EDUCATIONAL PSYCHOLOGY

continued from page ii

Joel R. Levin and Vernon L. Allen (eds.). Cognitive Learning in Children: Theories and Strategies

Donald E. P. Smith and others. A Technology of Reading and Writing (in four volumes).

Vol. 1. *Learning to Read and Write: A Task Analysis (by Donald E. P. Smith)*

Vol. 2. *Criterion-Referenced Tests for Reading and Writing (by Judith M. Smith, Donald E. P. Smith, and James R. Brink)*

Vol. 3. *The Adaptive Classroom (by Donald E. P. Smith)*

Vol. 4. *Designing Instructional Tasks (by Judith M. Smith)*

Phillip S. Strain, Thomas P. Cooke, and Tony Apolloni. Teaching Exceptional Children: Assessing and Modifying Social Behavior